Nonlinear Process Control:
Applications of Generic Model Control

Peter L. Lee (Ed.)

Nonlinear Process Control:
Applications of Generic Model Control

With 77 Figures

Springer-Verlag
London Berlin Heidelberg New York
Paris Tokyo Hong Kong
Barcelona Budapest

Peter L. Lee
Head, Department of Chemical Engineering
The University of Queensland
4072 Queensland
Australia

ISBN-13:978-1-4471-2081-0   e-ISBN-13:978-1-4471-2079-7
DOI: 10.1007/978-1-4471-2079-7

British Library Cataloguing in Publication Data
A catalogue record for this book is available from the British Library

Library of Congress Cataloging-in-Publication Data
A catalog record for this book is available from the Library of Congress

Typesetting: Camera ready by authors

69/3830—543210 Printed on acid-free paper

SERIES EDITORS' FOREWORD

The series *Advances in Industrial Control* aims to report and encourage technology transfer in control engineering. The rapid development of control technology impacts all areas of the control discipline. New theory, new controllers, actuators, sensors, new industrial processes, computing methods, new applications, new philosophies, . . ., new challenges. Much of this development work resides in industrial reports, feasibility study papers and the reports of advanced collaborative projects. The series offers an opportunity for researchers to present an extended exposition of such new work in all aspects of industrial control for wider and rapid dissemination.

Nonlinear control theory has a long history reaching back to the stability theory of nonlinear systems developed by mathematicians like A.M. Lyapunov. What has been less successful has been the application of nonlinear methodologies in industrial applications. This volume in the Advances in Industrial Control series is a valuable contribution to the task of rectifying this situation. Dr Peter Lee has drawn together a number of industrial applications of the nonlinear control technique variously known as Generic Model Control (GMC) or Reference System Synthesis (RSS) or Nonlinear Process Model Based Control (PMBC). The thrust of the technique is to match the simplicity of PI control tuning with the knowledge which resides in existing nonlinear models of industrial processes. The contributions in the volume are interesting and extremely instructive. For not only are the technical problems of using GMC covered but the contributors cover issues like how to obtain operator acceptance of new untried techniques, and how to persuade management to invest in new solutions. Equally instructive are the financial payback figures given which result from the investment in new advanced control schemes.

We are grateful to Dr Peter Lee and his co-authors for producing such a thought provoking contribution to the Advances in Industrial Control series.

Michael Grimble and Michael Johnson,
Industrial Control Centre,
University of Strathclyde,
Glasgow, Scotland, U.K.

PREFACE

Much has been written and said about the "theory and practice" gap that exists in Process Control. Academic control theory has not been well utilized in many industries. The big strides that were made in the 1960's in 'Modern State-Space' methods and to a lesser extent, in the 1970's on Adaptive Control, were largely ignored by industry in general. The PID regulator continues to meet the needs of many practitioners. However, even though the number of difficult control loops that cannot be solved using a PID regulator is small, the economic impact of poor control in these loops can still be quite large.

This book clearly is aimed at practitioners. It is not meant to be a full exposition of nonlinear control methods in all its guises. I have attempted, rather, to gather a series of case studies that illustrate important aspects of implementing advanced control in an industrial setting. Only one nonlinear control method is chosen for examination – Generic Model Control. No claim is made in this book that this technique is the best of the available nonlinear methods (although this author would claim it does have some considerable advantages!) No comparisons are made in this book of alternative control strategies for that is not the point. It is hoped that this book will demonstrate that advanced control methods can successfully be implemented and that the economic returns for doing so can be substantial.

The editor, Dr Peter Lee would like to take this opportunity to thank a number of people who have contributed to this monograph.

Firstly to the contributors who have endured my bullying and cajoling in equal proporitions, and who really made this work possible, a big thanks is due – "goodonya mates". To their respective companies who granted permission for the works to be published, we are all grateful for their contribution to the profession.

To my colleagues at the CAPE Centre, Dr Bob Newell and Dr Ian Cameron who put up with me and carried some extra load while I was writing, a big thank-you.

To Christine Smith who most of all endured the drafts, revisions and vagaries of the word processing graphics systems can only go my sincere thanks and much appreciation for her dedication to quality.

Finally to my family Janet and Geoffrey who have foregone many hours of leisure because "Daddy is working", my love and appreciation.

Peter Lee
Brisbane, 1993

CONTENTS

CONTRIBUTORS

Chapter 3: An Industrial Application of Reference System Synthesis Generic Model Control: Wastewater pH Control
R. Donald Bartusiak,
Exxon Chemical Company, Texas, USA

Chapter 4: Using Tray-To-Tray Models for Distillation Control
James B. Riggs[1], Martin Beauford[2] and Jackie Watts[2]
[1]Department of Chemical Engineering, Texas Tech University, Texas, USA
[2]Phillips 66 Company, Borger, Texas, USA

Chapter 5: Automatic Moisture Control in Particulate Dryers
D. Doerr[2], P.L. Douglas[1] and M.G. Whaley[2]
[1]Department of Chemical Engineering, University of Waterloo, Waterloo, Canada
[2]Dantec Systems Corporation, Waterloo, Ontario, Canada

Chapter 6: Eccentricity Control of a Cable Jacketing
B.W. Surgenor
Department of Mechanical Engineering, Queen's University, Kingston, Canada

Chapter 7: The Development of a Nonlinear Adaptive Generic Model Controller for Chemical Reaction Quality Control
Barry J. Cott,
Sarnia Manufacturing Centre, Shell Canada Limited, Ontario, Canada

Chapter 8: Blast Furnace Stove Control
G.A. Labossiere and P.L. Lee
Computer Aided Process Engineering Centre, Department of Chemical Engineering, The University of Queensland, Queensland, Australia

CHAPTER 1

INTRODUCTION

The need for high performance control systems has accelerated over the past decade. This need has been pushed by economic pressures related to increased throughput, higher quality products produced more consistently, increased utilization of raw materials and decreased energy utilization. The rise of environmental and safety issues and a tighter material and energy integration has also added to the demands of modern control systems. At the same time, increased computational power through modern distributed control systems has provided the means to implement many advanced control algorithms at reasonable costs.

While the need for high performance control systems is well known to most practitioners, less known is the means to quantify the benefits of implementing Advanced Process Control (APC). The Warren Centre Study (Marlin et al 1987) examined a series of case-studies and documented a methodology for justifying the expenditure required to implement APC. Despite this excellent work, one reason often cited by industrial practitioners for not implementing APC is the inability to cite other successful implementations. It is hoped that this monograph may in some small way alleviate this problem.

Most advanced algorithms rely, either implicitly or explicitly, on a process model to perform the control calculations. The development of process control algorithms has in the past taken a very different approach in using process models than other process systems related activities, including process design and optimization. The latter activities have made use of nonlinear, multivariable models, often based upon a physical understanding of the process behaviour. These models have often been collated and coded in standard simulation libraries such as ASPEN (Evans et al 1979) and others. Of course, these models have predominantly only considered the steady-state behaviour of the process. In contrast, process control has relied primarily on linear, single-input, single-output models, often derived from empirical plant tests. That this approach has failed to yield adequate results in all cases is hardly surprising. Most process plants behave in a nonlinear way, exhibiting interactions between many of the process variables. Thus, increasingly there is interest in developing methods that deal directly with these issues.

Generic Model Control (GMC)(Lee and Sullivan, 1988) was developed with the specific objective of incorporating nonlinear, multivariable process models directly in the control algorithm. Independently, similar approaches were developed by Bartusiak et al (1988) called Reference System Control (RSC) and by Balchen et al (1988) called Internal Decoupling. These approaches are all similar and can be traced to some earlier work by Liu (1967). They are also part of a distinct body of mathematical knowledge known as

2

differential geometry. One of the key differences in these approaches is the way in which performance is defined. While attaining the required setpoint for each controlled variable is still maintained as necessary, an additional requirement is also added. This additional requirement specifies the rate of approach towards the setpoint and it is this specification that differentiates these methods from other techniques.

This monograph will introduce the required "theory" in an attempt to convey the basic principles of GMC. This is written in a tutorial manner to help to explain the basic concepts of the approach. Details of the underlying theory and extensions will be left to the reader to pursue, if interested, through the cited references. A series of case studies describing the application of GMC to a number of processes then follows. These case studies highlight implementation aspects as well as the benefits of applying this technique to the particular problem described. Finally some overall conclusions are drawn in the final chapter.

The case studies presented in this monograph have been chosen to cover a wide range of different applications across a range of industries. The processes chosen also exhibit a variety of characteristics that make process control difficult. Table 1.1 summarizes the characteristics of each case study.

Table 1.1 Case Study Characteristics

Chapter	Case Study	Input/ Output Dimensions	Deadtimes Present	Constraints Present
3	pH Control	1 x 1	NO	NO
4	Distillation Control	2 x 2	NO	YES
5	Dryer Control	1 x 1	NO	NO
6	Extruder Control	2 x 2	NO	NO
7	Reactor Control	1 x 1	YES	NO
8	Stove Control	2 x 2	NO	YES

In this monograph the terms Generic Model Control (GMC), Reference Systems Synthesis (RSS), and Process Model Based Control (PMBC) will be used interchangeably.

1.1 REFERENCES

Balchen J.G., Lie B. and Solberg I. (1988) Internal decoupling in nonlinear process control. Modelling, Identif. and Control 9:137-148

Bartusiak R.D., Georgakis C. and Reilly M.J. (1988) Designing nonlinear control structures by reference system synthesis. Proc. ACC, Atlanta GA USA, 1585-1590.

Evans L.B., Boston J.F., Britt H.I., Gallier P.W., Gupta P.K., Joseph B., Mahalec V., Ng E., Seider W.D. and Yagi H. (1979) ASPEN: An advanced system for process engineering. Comput. & Chem. Eng. 3:319-327.

Lee P.L. and Sullivan G.R. (1988) Generic Model Control (GMC). Comput. & Chem. Eng. 12:573-580

Liu S-L. (1967) Noninteracting process control. Ind. Eng. Chem. Proc. Des. and Dev. 6:460-468

Marlin T.E., Perkins J.D., Barton G.W. and Brisk M.L. (1987) Advanced Process Control. The Warren Centre for Advanced Engineering, University of Sydney, Sydney, Australia.

CHAPTER 2

GENERIC MODEL CONTROL - THE BASICS

2.1 INTRODUCTION

The basic concept of Generic Model Control (GMC), (Lee and Sullivan, 1988), and its closely aligned cousins Reference System Synthesis (Bartusiak et al, 1988) and internal decoupling (Balchen et al, 1988) is to find values of the manipulated inputs that force a model of the system to follow a desired reference system or trajectory. The methods are clearly related to a body of mathematical knowledge known as differential geometry involving exact linearization of nonlinear mappings between the input and output variables (Isidori, 1989). The purpose of this chapter is not to present all of the theoretical underpinnings of the technique, but rather to provide enough of the essential elements of the method in a tutorial style that will allow the reader to appreciate the applications of the technique presented in each of the subsequent case studies.

2.2 THE BASICS

Consider a dynamic model of a process described by a set of differential equations:

$$\dot{\mathbf{y}} = \mathbf{f}(\mathbf{y},\mathbf{u},\mathbf{d},t,\theta) \tag{2.1}$$

where $\mathbf{y}$ is a vector of process outputs of dimension m, $\mathbf{u}$ is a vector of process inputs of dimension m, $\mathbf{d}$ is a vector of process disturbances of dimension l, t is time and θ is a vector of model parameters of dimension q. In general $\mathbf{f}$ is a vector of nonlinear known functional relationships. Note also in this simplified presentation we will only consider square systems - i.e. where the number of inputs and outputs are the same. However, the technique is not limited to only such systems (Lee and Sullivan, 1988).

The second component of the algorithm is to define a reference system, $\mathbf{r}(\mathbf{y})$. This reference system defines a desirable rate of change of the output variables, $(\dot{\mathbf{y}})^*$. This notation is consistent with the notation that defines the setpoint of $\mathbf{y}$ as $\mathbf{y}^*$.

One reasonable choice of the reference system $\mathbf{r}(\mathbf{y})$ is:

$$\mathbf{r}(\mathbf{y}) = (\dot{\mathbf{y}})^* = \mathbf{K}_1(\mathbf{y}^* - \mathbf{y}) + \mathbf{K}_2 \int_0^{t_k} (\mathbf{y}^* - \mathbf{y})\,dt \tag{2.2}$$

where t_k is the current instant of time.
This equation expresses the desires that:

8

(a)   When the system is a long way from setpoint, we would like the system to be travelling towards the setpoint quickly.

(b)   We would like to ensure offset free performance and thus if the system has been away from setpoint for some time we would like it to start moving more quickly towards setpoint.

The values of K_1 and K_2 can be chosen to specify a range of different process behaviours.   Their choice will be discussed in section 2.3.   Bartusiak et al (1989) examined other possible choices for the reference system $r(y)$.

The third and final element of the control algorithm is to ensure that the rate of change of the outputs follows the desired reference system.   Thus:

$$\dot{y} = (\dot{y})^* \tag{2.3}$$

Using equations 2.1 and 2.2 yields:

$$f(y,u,d,t,\theta) = K_1(y^*-y) + K_2 \int_0^t (y^*-y)dt \tag{2.4}$$

The control law to be solved at every sample time, equation 2.4, for the manipulated inputs, u, is a set of nonlinear, algebraic equations in the unknown variables.

A simple single-input, single-output problem may help to clarify the control law.   Consider a simple liquid level control problem shown in Figure 2.1. The control problem is to control the liquid level by adjusting the inlet flowrate M_1. A disturbance inflow M_2 also exists.

9

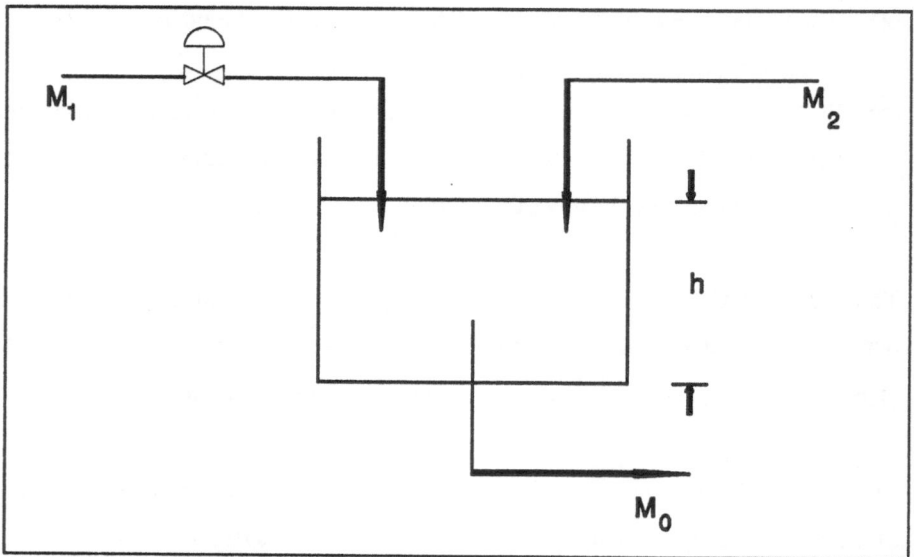

Figure 2.1 Gravity Underflow Tank

A simple model for this process is

$$\frac{dh}{dt} = M_1 + M_2 - C\sqrt{h} \qquad (2.5a)$$

since

$$M_0 = C\sqrt{h} \qquad (2.5b)$$

where h is the height of liquid in the tank and C is a constant. Comparing this specific model with the general case shown in equation 2.1, it is seen that

h	$\equiv$	y, the output variable
M_1	$\equiv$	u, the input variable
M_2	$\equiv$	d, the disturbance variable
and C	$\equiv$	θ, a model parameter.

For this system r(y) is given by:

$$(\dot{h})^* = K_1(h^* - h) + K_2 \int_0^t (h^* - h)dt \tag{2.6}$$

where h^* is the setpoint for h.

Using equations 2.4, 2.5 and 2.6 yields:

$$M_1 + M_2 - C\sqrt{h} = K_1(h^* - h) + K_2 \int_0^t (h^* - h)dt \tag{2.7}$$

or

$$M_1 = -M_2 + C\sqrt{h} + K_1(h^* - h) + K_2 \int_0^t (h^* - h)dt \tag{2.8}$$

In this case the control law given by equation 2.8 is an explicit expression for the manipulated input M_1 and is shown in Figure 2.2. Sometimes however, it is not possible to algebraically manipulate equations 2.4 to obtain such explicit expressions. In these cases equation 2.4 must be solved numerically. However, the task of solving the nonlinear simultaneous equations is often very fast, only requiring one or two iterations of a numerical method. This is because you can provide such a good initial guess for the solution - the values of the manipulated inputs at the previous sample time.

The derived control law in equation 2.8 represents a feedback/feedforward controller. The feedforward component would require measurement or estimation of the disturbance flowrate M_2. The control law is also dependent upon the model parameter C. Signal and Lee (1992) show how these disturbances and parameters may be estimated on-line when their values are not known. Much of the work on GMC has been to demonstrate how "robust" the method is when such disturbances, parameters and other model inaccuracies are present (Lee, 1991). An illustration of this property will be given in an example at the end of section 2.3, and brief summary of the major results given in section 2.6.

A final comment on the basic algorithm is required. A solution to equation 2.4 is only directly possible when the manipulated variable chosen to control a particular output appears in the model equation for that output. These types of systems are known as relative degree one. Other approaches can be taken when this is not the case (Lee, 1991; Henson and Seborg, 1990; Signal, 1992).

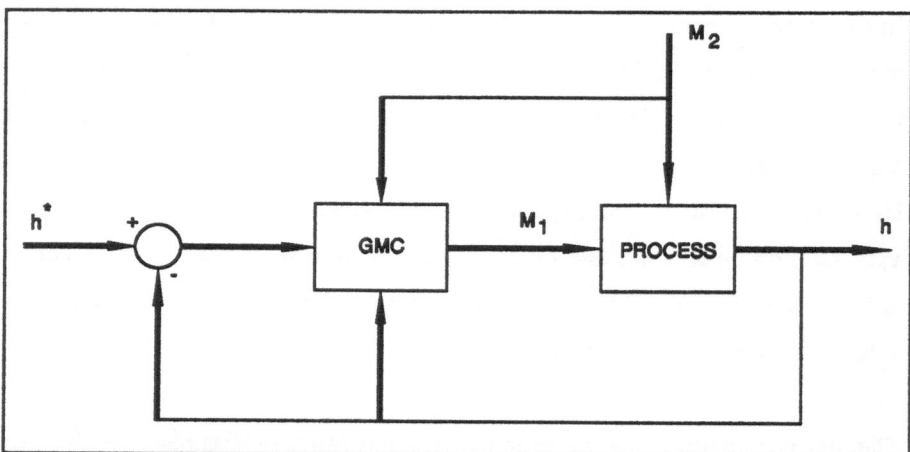

Figure 2.2 Block Diagram of Control Structure

2.3 REFERENCE SYSTEMS

As discussed in section 2.2, the reference system is one of the key elements in the GMC control law. The form given in equation 2.2 is only one of many possible choices (Bartusiak et al, 1989), and contains two parameters K_1 and K_2. These are diagonal matrices, where the diagonal elements are chosen for each output independently. Thus, one can use these values to select any "reasonable" desired response for the system.   "Reasonable" implies that the parameters are chosen in relation to the system's natural dynamic response. How well the system matches this reference system will be governed by how closely the chosen model matches the plant behaviour.

If Laplace transforms of equation (2.2) are taken for a single-input, single output system, the resulting transfer function becomes:

$$\frac{Y(s)}{Y^{*}(s)} = \frac{2\tau\xi s + 1}{\tau^2 s^2 + 2\tau\xi s + 1} \tag{2.9}$$

where

$$\tau = \frac{1}{\sqrt{k_2}} \quad \text{and} \quad \xi = \frac{k_1}{2\sqrt{k_2}}$$

This system does not yield the same response as a classic second-order system (Stephanopoulos p. 188, 1984), due to the presence of the zero in the transfer function. However, similar plots to the classic second-order response showing the normalized response of the system y/y^{*} vs normalized time t/τ with ξ as a parameter can be produced as is shown in Figure 2.2. The design procedure can be specified as follows:

13

1.   Choose ξ from Figure 2.3 to give desired shape of response.

2.   Choose τ from Figure 2.3 to give "appropriate" timing of response in relation to known or estimated plant speed of response.

Calculate k_1 and k_2 using the following equations:

$$k_1 = \frac{2\xi}{\tau} \quad , \tag{2.10}$$

$$k_2 = \frac{1}{\tau^2} \tag{2.11}$$

An alternative approach is to examine the characteristic equation of the closed-loop system. From equation 2.9, the characteristic equation is given by:

$$s^2 + k_1 s + k_2 = 0 \tag{2.12}$$

It may be desirable or more natural to specify the location of the two closed loop poles and then solve equation 2.12 for the unknown values of k_1 and k_2.

To bring some of these concepts into focus consider the control of an anaerobic digestor.

An anaerobic digestor is a process of converting organic material in a liquid stream, by a series of biological reactions, to methane and carbon-dioxide gas, and an effluent stream of less concentration. It is commonly used in treating waste-water streams.

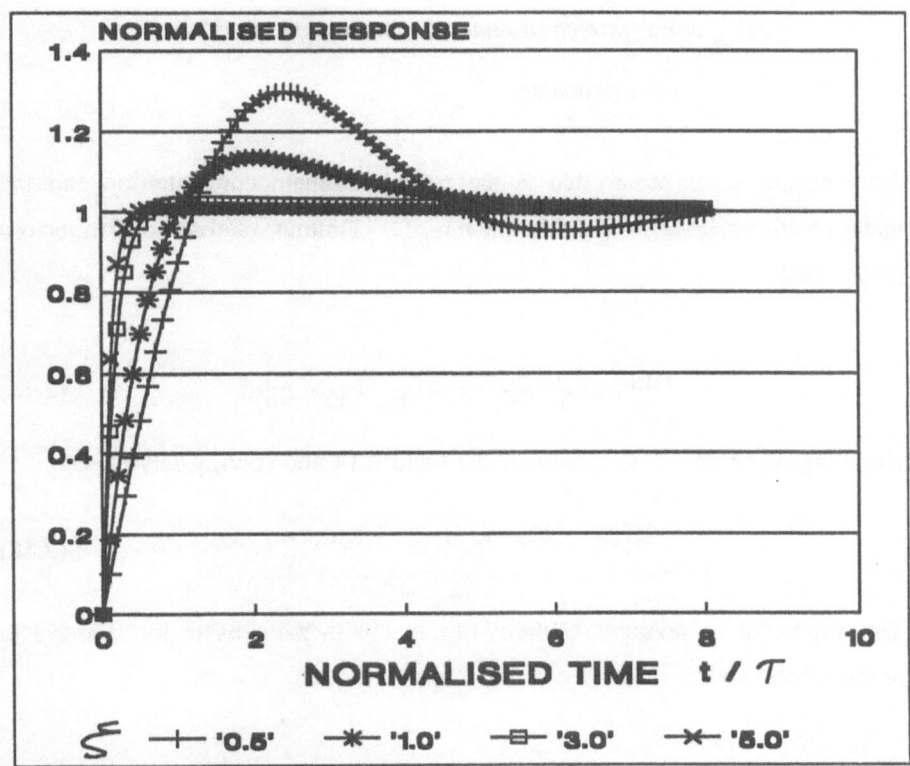

Figure 2.3 Reference System Curves

Consider a simplified model of anaerobic digestor described by Costello et al (1989b):

$$\frac{dS}{dt} = D(S_{in} - S) - \alpha_1 Q_{CH_4} \qquad (2.13)$$

where   S   is the organic effluent discharge concentration
                  (g COD/dm³)

           S_{in} is the organic influent concentration (g COD/dm³)

           D   is the dilution rate (hr⁻¹) (inlet flowrate/volume of reactor)

15

Q_{CH_4} is the production rate of methane gas (dm^3/hr)

α_1 yield parameter

The variable to be controlled is the organic effluent concentration and the manipulated variable is the dilution rate, D. Defining a reference trajectory for S yields

$$\left(\frac{dS}{dt}\right)^* = K_1 (S^* - S) + K_2 \int_0^{t_s} (S^* - S)dt \qquad (2.14)$$

Applying equation 2.3 to equations 2.13 and 2.14 and solving D yields

$$D = \frac{K_1(S^* - S) + K_2 \int_0^{t_s} (S^* - S)dt + \alpha_1 Q_{CH_4}}{S_{IN} - S} \qquad (2.15)$$

This represents a nonlinear control law that will compensate for changes in process behaviour.

The control law of equation 2.15 and the system described in equation 2.13 were simulated using MATLAB (1990). The script file for this simulation is shown in Figure 2.4. The values of the simulation constants S_{IN}, α, and Q_{CH_4} were those used previously by Costello et al (1989b) and are representative of those found in some industrial processes. The value of the GMC reference trajectory parameter τ was chosen such that closed loop time constant should be 9 hours (compared with the open-loop time constant of 21 hours). This was chosen to illustrate the effective improvement that can be obtained using the GMC algorithm. The value of the other reference trajectory parameter ξ

16

was chosen on the basis that some overshoot in this process is acceptable.

Figure 2.5 shows the result of controlling the process using "perfect control". A step change in the inlet concentration, S_{IN}, was introduced after 10 hours. The controller, equation 2.15, knew of this change, contained no parameter errors, and hence maintained perfect control.

Figure 2.6 shows what occurs when no feedforward action is included in the control law. In this instance, the value of the inlet concentration, S_{IN}, used in the control law was always constant at 30 $gCOD/dm^3$, while the actual inlet concentration changed to 60 $gCOD/dm^3$ after 10 hours.

Figure 2.7 shows what occurs when parameter mismatch exists in the control law. The value of the parameter, α_1, used in the control law, was 50% larger than the value in the "plant". The same disturbance as was used previously was again applied, but in this instance it was assumed that the controller did have feedforward information. Notice also that the controller takes action from time zero as the controller is no longer initialized correctly.

17

```
% GMC Control of an Anaerobic Process
%
% Define Some constants
%
clear
simtim = 80.0;                  % Simulation Time
sin = 30.0;                     % Inlet Substrate Concentration
alpha1 = 1.0;
qch4 = 1.0;
tau = 9.0;                      % Reference Trajectory Specification
xi = 1.0;
k1 = 2.0 * xi / tau;
k2 = 1.0 / tau / tau;
s = 9.0;                        % Initial values
d = 0.0476;
setpoint = 9.0;
esum = 0.0;
splt = [splt ; s];
dplt = [dplt ; d];
%
% Main simulation loop
%
for i = 1:simtim
%
% Calculate Control action
%
    if i > 10
        sin = 60;
    end
    err = setpoint - s;
    esum = esum + err;
    reftraj = k1 * err + k2 * esum;
    d = (reftraj + alpha1 * qch4) / (sin - s);
    if d < 0
        d = 0.0;
    end
%
% now simulate process response
%
    dsdt = d*(sin - s) - alpha1*qch4;
    s = s + dsdt * 1.0;
%
% form plotting arrays
%
  splt = [splt ; s];
  dplt = [dplt ; d];
  ds = [ds ; dsdt];
end
subplot (211), plot (splt), title('s');
subplot (212), plot (dplt), title('d');
subplot
```

Figure 2.4 MATLAB script file for Anaerobic System

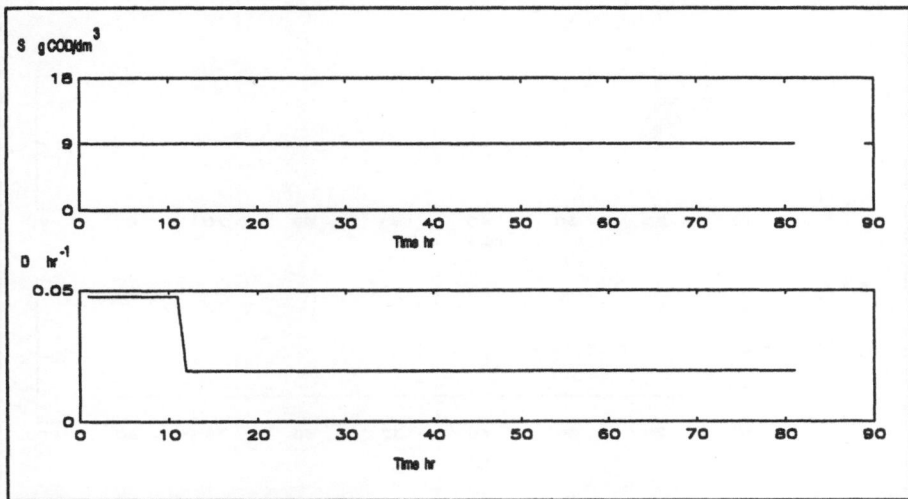

Figure 2.5　"Perfect" Control

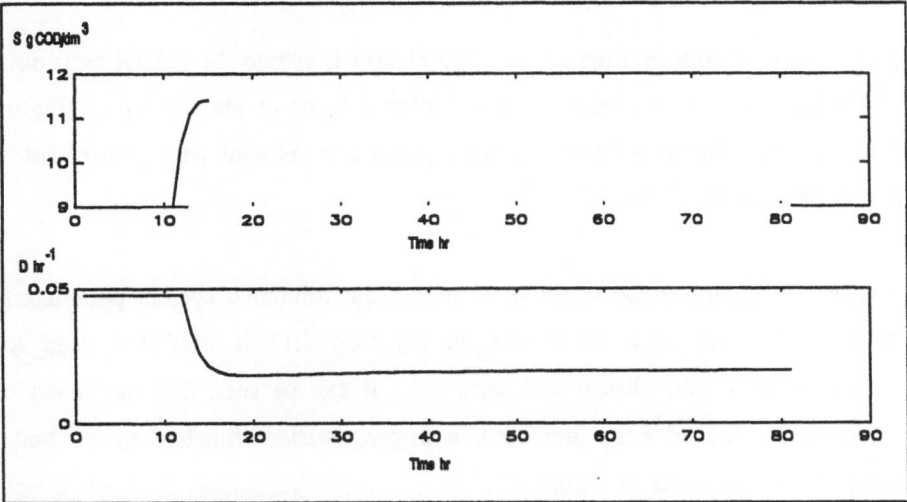

Figure 2.6　No Feedforward Control

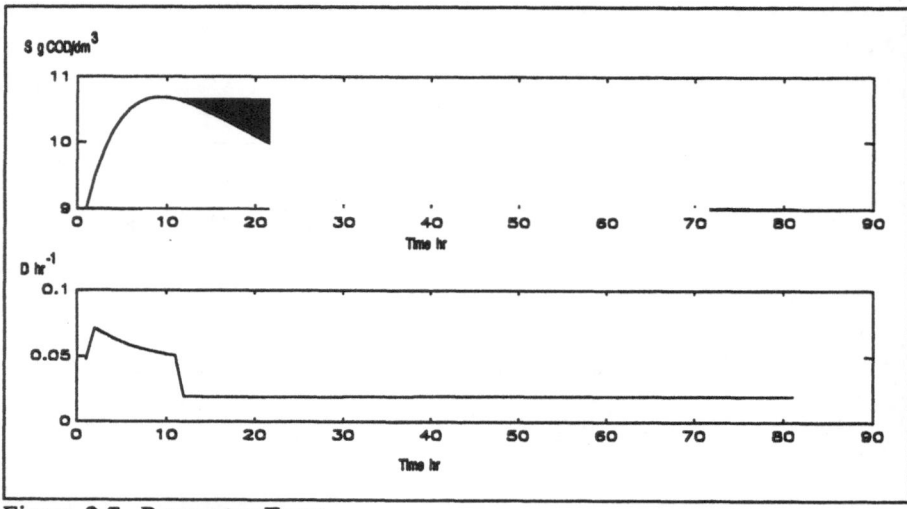

Figure 2.7   Parameter Error

The results of both Figures 2.6 and 2.7 clearly illustrate the robust behaviour of GMC. Despite modelling errors (Figure 2.6) or parameter errors (Figure 2.7), the controller is able to return the process to setpoint with zero offset in a smooth manner.

Figure 2.8 illustrates the effect of changing the reference system parameters. In this case, the value of τ used in equation 2.11 is half that used for obtaining the results shown in Figure 2.7. It can be seen that the speed of response has indeed been increased, with the transient finished by 40 hours compared to 80 hours in Figure 2.7.

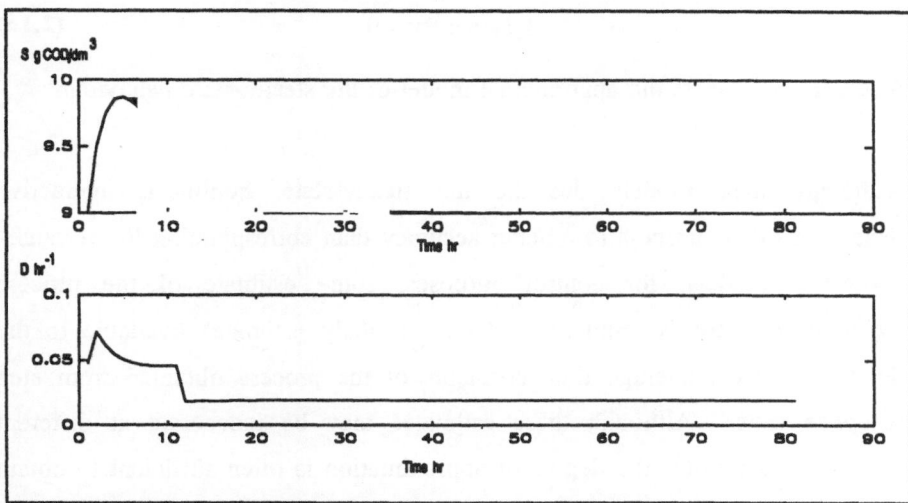

Figure 2.8 Effect of Reference System Parameters

2.4 USING STEADY STATE MODELS

The basic algorithm as described in section two has been extended to include the use of steady-state models, systems with dead-time, coping with model uncertainty and controlling systems in the presence of constraints. A brief review of using steady-state models is given in this section, and an extended development is to be found in Cott et al (1989).

The process model of equation 2.1 assumes an approximate dynamic model of the process can be derived, but it is far more common to have access to models that only contain steady-state information, i.e.

21

$$\mathbf{f_{ss}(y,u,d,\theta)} = 0 \qquad (2.16)$$

where $\mathbf{f_{ss}}$ represents the approximate model of the steady-state behaviour.

Although these models describe the steady-state, nonlinear, interactive behaviour of the process to a better accuracy than corresponding linear cause-and-effect models, for control purposes, some estimate of the process dynamic response is required. The most likely estimates available to the designer are the average time constants of the process obtained from step response tests. Although these estimates may be inaccurate at different operating conditions, the degree of approximation is often sufficient to obtain good control performance.

Assuming that the step response data can be represented by a first-order response, a simple estimate of the time response of the output variables in moving from one steady-state to another can be given as:

$$\mathbf{\dot{y}} \approx \mathbf{T^{-1}(y_u - y)}, \qquad (2.17)$$

where $\mathbf{T}$ is a diagonal matrix of the estimated open-loop time constants, and $\mathbf{y_u}$ are the ultimate or steady-state values of the output variables if no further control action is taken. The diagonal elements of the matrix $\mathbf{T}$ are "averaged" time constants of the output variables based on step changes of all input variables. Combining this approximate description of the dynamic response of the process with the reference system or performance specification of equation 2.2, the ultimate response can be calculated as:

$$y_u = y + T(K_1(y^* - y) + K_2 \int_0^t (y^* - y)dt) \qquad (2.18)$$

The control action required to achieve this performance specification can be found by replacing y with y_u in the nonlinear, steady-state model described in equation 2.16 and solving for the control variables u. An example will clarify this procedure.

Consider the problem of controlling the anaerobic digestor described in the previous section. Imagine that only a steady-state mass balance is available to describe this system:

$$D(S_{IN} - S) - \alpha_1 Q_{CH_4} = 0 \qquad (2.19)$$

This equation is in the form of equation 2.16. Applying equation 2.18 for the output variable S yields:

$$S_u = S + \tau_p(K_1(S^* - S) + K_2 \int_0^t (y^* - y)dt) \qquad (2.20)$$

where τ_p is an estimate of the open loop time constant. Using equation 2.20 to determine the ultimate value of the substrate concentration if no further control action is taken, equation 2.19 can be used to calculate the manipulated variable required to achieve S_u by substituting S_u for S:

$$D = \frac{\alpha_1 Q_{CH_4}}{(S_{IN} - S_u)} \qquad (2.21)$$

Equations 2.20 and 2.21 form the complete control algorithm. Note that if:

$$\tau\, K_1 = 1$$
and $\quad K_2 = 0$
then $\quad S_u = S^*$.

Thus the control law will result in calculating a value of D that according to the steady-state model should give the required setpoint. Equation 2.20 will compensate for model errors and give improved dynamic performance.

It should be pointed out that the use of equation 2.17 does not require the steady-state model to be explicit in either output or manipulated variables, making the availability of suitable models to develop control laws much greater than if explicit models were required. Depending on the model structure and number of control variables, solving for the control variables may be straightforward, or may require numerical methods for solving systems of nonlinear equations.

Implicit steady-state process models, such as those described in equation 2.16 do not model the dynamic effects of the disturbance variables **d**. A simple, but effective method to improve the model's performance is to pass the disturbance variables through appropriate dynamic filters (eg. first-order lags) to compensate for the lack of dynamic structure in the process model. These filters need to be tuned on-line to achieve good control performance. This is similar to conventional feedforward control.

24

2.5 CONSTRAINT HANDLING STRATEGY

The control law derived in section 2.2 was derived by <u>equating</u> equations 2.1 and 2.2. Equally however, a control law could be developed by posing the problem as an optimization problem (Lee and Sullivan, 1988). This optimization problem would seek to minimize the difference between the process and reference system by choosing specific values of the manipulated variables. Of course <u>equating</u> is certainly minimizing. However, when constraints on the inputs and outputs are present, a constrained optimization approach is more suitable. A method where slack variables defining the variables departure from the chosen reference trajectories are added to the control law for both the control variables and constraint variables has been proposed (Brown et al, 1989). Selecting the weighting factors on these slack variables and defining a control objective function which is dependent on these weighted slack variables allows the controller to achieve the desired compromise between constraint violation and setpoint tracking. The solution becomes a nonlinear constrained optimization problem.

Consider the process described by equation 2.1 where the number of outputs and the number of inputs are equal.

It is desirable to have the system operate within the feasible region such that for the q constraints:

$$C_{Li} \leq \psi_i(\mathbf{y}, \mathbf{u}, \mathbf{d}, t, \theta) \leq C_{ui} \; ; \quad i = 1 \; ... \; q \tag{2.22}$$

where ψ_i represents some nonlinear function such that:

25

$$C_i = \psi_i(\mathbf{y},\mathbf{u},\mathbf{d},t,\theta) \; ; \;\; i=1 \ldots q$$

As well, both input constraints and input movement constraints are defined for the m controls, such that:

$$u_{Li} \leq u_i \leq u_{Ui} \; ; \;\; i=1 \ldots m \tag{2.23}$$

$$\Delta u_{Li} \leq u_i(t+\Delta t) - u_i(t) \leq \Delta u_{Ui} \; ; \;\; i=1 \ldots m \tag{2.24}$$

If the output of the process is such that it lies outside the feasible region or that the current output variable trajectories will violate the given process constraints, then it is desirable to operate the system such that the rate of change of the constraint variable C_i approaches its constraint value according to the reference trajectory:

$$\frac{dC_i}{dt} = K_{1Ci}(C_{Ui} - C_i) \; ; \;\; i=1 \ldots q \tag{2.25}$$

or

$$\frac{dC_i}{dt} = K_{1Ci}(C_i - C_{Li}) \; ; \;\; i=1 \ldots q \tag{2.26}$$

where K_{1Ci} are chosen a-priori using the GMC reference trajectories.

The dynamics of the constraint variables can be found by utilizing the chain rule on equation 2.22. Alternatively, an approximation like equation 2.17 would be to assume that the constraints will move to their constraint value in a first order manner according to:

$$\frac{dC_i}{dt} = T_{Ci}^{-1}(C_i^{AIM} - C_i) \; ; \;\; i=1 \ldots q \tag{2.27}$$

$$C_i^{AIM} = \psi_i(y^{AIM}, \mathbf{u}, \mathbf{d}, t, \theta) \quad ; \quad i = 1 \ldots q \tag{2.28}$$

$$f(y^{AIM}, \mathbf{u}, \mathbf{d}, t, \theta) = 0 \tag{2.29}$$

The q slack variables, λ_{ci}^- and λ_{ci}^+, defining the variables departure from the chosen specification curves are added to equations 2.25 and 2.26 for the constraint variables. For the case of the known constraint dynamics, equations 2.25 and 2.26 become:

$$\frac{dC_i}{dt} - \lambda_{ci}^- \leq K_{1Ci}(C_{Ui} - C_i) \quad ; \quad i = 1 \ldots q \tag{2.30}$$

and

$$\frac{dC_i}{dt} + \lambda_{ci}^+ \leq K_{1Ci}(C_i - C_{Li}) \quad ; \quad i = 1 \ldots q \tag{2.31}$$

where λ_{ci}^- and λ_{ci}^+ represent the variables departure from the chosen reference trajectories for the upper and lower constraints respectively. Defining a reference system for the constraint variables not only allows the system to ensure that the constraint is not violated, but provides control over the rate of approach to the constraint through the selection of K_{1Ci}.

A set of slack variables can also be incorporated into the performance curves to denote the systems efficiency in terms of setpoint tracking. If two mutually independent slack variables, λ_{pi}^- and λ_{pi}^+, are defined to express the systems negative offset and positive offset from the prespecified response trajectory, the GMC control law for the system performance in equation 2.4 can be written as:

$$f(y,u,d,t,\theta) + \lambda_{pi}^{+} - \lambda_{pi}^{-} = K_{1i}(y_i^{*} - y_i) + K_{2i} \int_{0}^{t} (y_i^{*} - y_i)\,dt \; ; \; i = 1\,...m \qquad (2.32)$$

where y_i^{*} represents the process setpoints.

The solution of the constrained multivariable control problem can be solved as a nonlinear constrained optimization problem, which minimizes a function of the slack variables.

The overall control problem can be formulated as the following single step nonlinear optimization problem:

NLP1:

Choose: $u, \lambda_{pi}^{+}, \lambda_{pi}^{-}, \lambda_{ci}^{+}, \lambda_{ci}^{-} \; ; \; i = 1\,...\,q$

To Minimize:

$$J = \omega_{pi}.(\lambda_{pi}^{+})^2 + \omega_{pi}.(\lambda_{pi}^{-})^2 + \omega_{cj}.(\lambda_{cj}^{+})^2 + \omega_{cj}.(\lambda_{cj}^{-})^2 + \omega_{\Delta ui}(\Delta u_i)^2 \begin{matrix} i = 1 & ...\,m \\ j = 1 & ...\,q \end{matrix} \quad (2.33)$$

where ω_i and ω_j are weights such that

$$\omega_i \geq 0$$

and

$$\omega_j \geq 0$$

Subject to:

$$f(y,u,d,t,\theta) + \lambda_{pi}^+ - \lambda_{pi}^- = K_{1i}(y_i^* - y_i) + K_{2i}\int_0^{t_i} (y_i^* - y_i)dt \; ; \; i=1\ldots m$$

$$(2.34)$$

$$\frac{dC_i}{dt} - \lambda_{ci}^- \leq K_{1Ci}(C_{Ui} - C_i) \; ; \; i=1\ldots q \qquad (2.35)$$

$$\frac{dC_i}{dt} + \lambda_{ci}^+ \leq K_{1Ci}(C_i - C_{Li}) \; ; \; i=1\ldots q \qquad (2.36)$$

$$C_i = \psi_i(y,u,d,t,\theta) \; ; \; i=1\ldots q \qquad (2.37)$$

$$u_{Li} \leq u_i \leq u_{Ui} \; ; \; i=1\ldots m \qquad (2.38)$$

$$\Delta u_{Li} \leq u(t+\Delta t)_i - u(t)_i \leq \Delta u_{Ui} \; ; \; i=1\ldots m \qquad (2.39)$$

$$\lambda_{pi}^+ \geq 0; \quad \lambda_{pi}^- \geq 0; \quad \lambda_{cj}^+ \geq 0; \quad \lambda_{cj}^- \geq 0 \; ; \begin{matrix} i=1\ldots m \\ j=1\ldots q \end{matrix} \qquad (2.40)$$

The overall problem described above can be solved as a single time step NLP.

The fact that separate control performance curves have been defined for the constraints as well as the controlled variables provides a great deal of flexibility in the controller design. The weighting terms are responsible for assigning a priority level to the various control objectives, while the

29

performance parameters are able to define the trajectories of the constraint paths.   Therefore, if the set point of a given control variable is increased above a heavily weighted upper variable constraint, then the control variable trajectory will deviate from its pre-defined performance curve to follow the trajectory defined for the constraint variable.   It is the degree of flexibility in establishing the proper balance between the violation of the constraint variables and the deterioration of the control performance, coupled with the ability to predefine the reference trajectories for both the controls and constraints, which are the merit of this technique.

The optimization problem which arises can be solved using a nonlinear constrained optimization algorithm.   The form of the optimization problem is well structured since a slack variable is added to each control law equation to ensure that a solution to the set of equations does in fact exist.   If the control is implemented at a reasonable frequency, the solution of the NLP is very fast (3 to 4 iterations) since the current control settings and slack variables provide a good initial estimate of the solution vector.

Brown et al. (1989) successfully demonstrated the application of this approach to the simulated forced-circulation evaporator.   The simulation study explored the effects of changes in the algorithm parameters and clearly demonstrated the efficiency and ease of the approach.

2.6 ROBUST STABILITY AND PERFORMANCE ANALYSIS

One of the most important properties of any system is closed-loop stability. Unfortunately the analysis of the stability of nonlinear closed-loop feedback systems is quite difficult, and requires advanced mathematical concepts and tools. Secondly, when a control algorithm relies upon a mathematical model, it is important to understand the effect of modelling errors on the closed-loop stability and performance. This is known as Robustness Analysis, an area also made difficult by advanced mathematical concepts.

The robust stability and performance analysis of GMC has been developed by Zhou et al (1992b) and by Signal (1992). These works provide a detailed analysis and only a summary of the key results will be presented here.

Both Zhou et al (1992b) and Signal (1992) show that:
1.   If no model errors exist and the model is minimum phase
2.   If no un-measured disturbances exist
3.   If no constraints on the manipulated variables exist
4.   If the sampling interval is insignificant compared to the process dynamics

then the closed loop system will be stable for any positive values of the reference trajectory parameters ξ and τ. This is the "Nominal" stability criterion and is quite restrictive in terms of the assumptions made. However, the proof involved the application of Lyapunov theory and the demonstration that a Lyapunov function could be found (Zhou et al, 1992b).

Signal (1992) analysed the closed-loop system by linearizing the nonlinear system about some nominal operating point. Obviously this approach cannot then make global claims about the stability and performance of the system, but rather claims that are valid in the region around the point of linearization.

Consider the linearized process model to be given by:

$$\dot{x} = \hat{A}x + \hat{B}u + \hat{D}d \qquad\qquad \textbf{(2.41a)}$$

$$y = \hat{C}x \qquad\qquad \textbf{(2.41b)}$$

and the real process as:

$$\dot{x} = Ax + Bu + Dd \qquad\qquad \textbf{(2.42a)}$$

$$y = Cx \qquad\qquad \textbf{(2.42b)}$$

The analysis also considered the presence of noise and unmeasured disturbances. The closed-loop system in the presence of modelling errors is stable if:

$$\mu_\Delta (W_1 T_N \, W_2) \leq 1 \qquad\qquad \textbf{(2.43)}$$

for "structured" uncertainties where

$$T_N = [sI + (sI - A)^{-1} \, B(\hat{C}\hat{B})^{-1} \, ((K1s + K2) \, \hat{C} + \hat{C}\hat{A}s)]^{-1}$$
$$(sI - A)^{-1} B(\hat{C}\hat{B})^{-1}((K1s + K2)\hat{C} + \hat{C}\hat{A}s) \qquad\qquad \textbf{(2.44)}$$

    s   is the Laplace Operator

K1 and *K2* are the GMC reference trajectory parameter values
arranged in diagonal matrices

W1 and *W2* are weighting matrices

and $\mu_\Delta(.)$ is the structured singular value.

Equation 2.43 (and hence 2.44) needs to be evaluated over a range of frequencies of interest. "Structured" uncertainties implies that the uncertainties in the model can be ascribed to specific model components rather than some overall accuracy statement.

Performance of the controller when modelling errors are present is also important. For any unmeasured disturbance D' that satisfies:

$$| W_I^{-1} D' |_2 \leq 1 \tag{2.45}$$

the weighted Integral Square Error of the outputs is guaranteed to be less than or equal to J where J is:

$$J = \underset{\Delta}{Max} | W_0 S_D' [I + (W_2 \Delta W_1)T_N]^{-1} W_I |_\infty \tag{2.46}$$

and $\sigma^* (\Delta) \leq 1$

where

$$S_{D'} = [I + s^{-1} (sI - \hat{A})^{-1}((K1 + \hat{A})s + K2)]^{-1} \tag{2.47}$$

Δ represents the structured model uncertainty

and $\sigma^*(.)$ is the maximum singular value

J is a function of the nominal model, its uncertainty description, the GMC reference trajectory parameters and two weighting matrices W_I and W_0. W_0 weights different outputs according to their importance in controlling the plant. W_I ensures that condition 2.45 holds.

Condition 2.46 can be used to determine the 'best' values of the GMC reference trajectory parameters. It was noted in the previous paragraph that the robust performance J is dependent on the GMC tuning parameters *K1* and *K2*. This is obvious for the nominal case when the model is perfect. For such perfect models, increasing the elements of *K1* and *K2* will result in improved performance. Increasing *K1* and *K2* can be achieved by increasing the GMC reference trajectory shape parameters ξ_i and decreasing the speed parameters τ_i. In the presence of model error, however, excessive demands placed on *K1* and *K2* may lead to inferior performance, or even instability. It is clear then, that an optimal performance J_{opt} can be found which minimises J with respect to *K1* and *K2*. i.e.:

$$J_{opt} = \min_{K1,K2} J \qquad (2.48)$$

J_{opt} is a function of the nominal model, its uncertainty description, the input weighting matrix W_I and the output weighting matrix W_0.

It is important therefore, that the uncertainty description is representative of the uncertainty associated with the model. Excessive or overly conservative descriptions will lead to lower *K1* and *K2* values than are necessary, resulting in sluggish control when implemented. Conversely, imprecise and narrow uncertainty descriptions will lead to higher *K1* and *K2* values, and this may result in inferior closed-loop performance. There is a need, therefore, to develop accurate and reliable methods for quantifying the uncertainty associated with various types of structural mismatch.

Examples of the use of both the robust stability and performance criteria can be found in Signal (1992). However, the theory recently developed has not had any impact of the applications presented in this book, or the applications described in the next section. In the future, the theory does offer the potential to guide model development as discussed by Signal (1992).

2.7 APPLICATIONS

GMC has been applied to a range of industrial, pilot-scale and simulated processes. A partial list of known applications has been compiled and is shown in the Table 2.1 (Lee, 1991). This table only lists "major" application studies where the results have either been applied industrially to pilot-plants or to extensive simulations involving many equations and variables. The table classifies the applications according to the type of model and process characteristics. In summary, these applications have all indicated the practicality of implementing the direct approach for a wide range of processes, and have demonstrated good control performance against the stated control objectives.

Table 2.1 GMC Applications

Application	Reference	Deadtimes Present	Constraints Present	Type of Model Used	Type of Environment
Anaerobic Digestion	Costello et al (1989a,1989b,1990a, 1990b,1993) Costello (1990) Greenfield et al (1990)	NO	NO	Steady-state and Dynamic	Simulation
Activated Sludge Plant	Lee et al (1988)	NO	NO	Dynamic	Simulation
Sugar Crystallizers	Wilson et al (1988) Wilson (1990)	NO	NO	Dynamic	Pilot Plant
Crude Tower	Zhou and Lee (1990)	YES	YES	Dynamic	Simulation
Evaporator	Lee et al (1989) Newell and Lee (1989)	NO	NO	Dynamic	Simulation
Blast Furnace Stoves	Labossiere (1990)	NO	NO	Steady-state and Dynamic	Pilot Plant
Blast Furnace	Zhou et al (1992a)	YES	NO	Steady-state	Industrial/ Simulation
Reactors	Brown (1989)	NO	NO	Dynamic	Industrial/ Simulation
	Cott and Macchieto (1989)	NO	NO	Dynamic	Simulation
	Howie (1988)	NO	NO	Dynamic	Industrial/ Simulation
Distillation	Cott et al (1989)	NO	NO	Steady-state	Industrial/ Simulation
	Fountain (1990)	NO	NO	Steady-state	Industrial/ Simulation
	Malik (1988)	NO	NO	Steady-state	Industrial
	Riggs (1989a)	NO	NO	Steady-state	Industrial
	Sinha & Riggs (1989)	NO	NO	Steady-state	Industrial
Pneumatic Positioner	Surgenor (1990)	NO	NO	Dynamic	Pilot Plant

Application	Reference	Deadtimes Present	Constraints Present	Type of Model Used	Type of Environment
Thermal/ Hydraulic Process	Liu et al (1990)	NO	NO	Dynamic	Simulation
Furnace	Surgenor and Hesketh (1989)	NO	NO	Dynamic	Simulation
Surge Tank	Lee et al (1991)	NO	YES	Dynamic	Simulation

2.8 REFERENCES

Balchen J.G., Lie B. and Solberg I. (1988) Internal decoupling in nonlinear process control. Modelling, Identif and Control 9:137-148

Bartusiak R.D., Georgakis C. and Reilly M. (1989) Nonlinear feedforward/feedback control structures designed by reference system synthesis. Chem. Eng. Sci. 44:1837-1851.

Bartusiak R.D., Georgakis C. and Reilly M.J. (1988) Designing nonlinear control structures by reference system synthesis. Proc. ACC, Atlanta, Ca, Session TPG-4:30: 1585-1590

Brown M.W. (1989) Operational and Control Policies of an Acetylene Hydrogenation Reactor, M.A.Sc. Thesis, University of Waterloo, Waterloo, Ontario, Canada

Brown M.W., Lee P.L., Sullivan G.R. and Zhou W. (1989) A Constrained Nonlinear Multivariable Control Algorithm, AIChE Annual Meeting, San Francisco, California, November 5-10.

Costello D.J., Greenfield P.F. and Lee P.L. (1989a) Cost Effective Operating Strategies and Control of High-Rate Anaerobic Reactors. Anaerobic Digestion Workshop, Massey University, New Zealand 13-14 November.

Costello D.J., Lee P.L. and Greenfield P.F. (1989b) Anaerobic Digestion Control by Generic Model Control. Bioproc. Eng 4:119-122.

Costello D.J. (1990) Modelling, Optimisation and Control of High Rrate Anaerobic Reactors, PhD Thesis, Department of Chemical Engineering, The University of Queensland, Queensland 4072, Australia.

Costello D.J., Lee P.L. and Greenfield P.F. (1990a) Application of Generic Model Control to the pH Control of a Single-Stage High-Rate Anaerobic Reactor. Int. Assoc. Water Pollution and Control (IAWPRC) Conference, Kyoto, Japan.

Costello D.J., Greenfield P.F. and Lee, P.L. (1990b) Strategies to Minimise the Operational Costs of High-Rate Anaerobic Treatment Systems. Regional Seminar on Management and Utilisation of Agricultural and Industrial Wastes, Universite Malaya, Kuala Lumpur, Malaya, 21-23 March.

Costello D.J., Lee P.L. and Greenfield P.F. (1993) Strategies for the pH Control of Two-Stage High-Rate Anaerobic Treatment Plants. Water Research (submitted).

Cott B.J., Durham R.G., Lee P.L. and Sullivan G.R. (1989) Process model based engineering. Comp & Chem Eng 13:973-984

Cott B.J. and Macchieto S. (1989) Temperature control of Exothermic batch reactors using Generic Model Control. Ind Eng Chem Res 28:1177-1184

Fountain P. (1990) Process Model-Based Control of High Purity Distillation - An Industrial Case Study, M.A.Sc. Thesis, University of Waterloo, Waterloo, Ontario, Canada.

Greenfield P.F., Nalbantoglu M., Costello D., Newell R.B. and Lee P.L. (1990) Improved Operation of Wastewater Treatment Plants: Combined Roles of Biotechnology and Information Processing. 4th WPCF/JSWA Joint Technical Seminar on Sewage Treatment Technology, Tokyo, May 17-19.

Henson M.A. and Seborg D.E. (1990) A critique of Differential Geometric Control Strategies for Process Control. IFAC World Congress, Tallinn, Estonia

Howie B.A. (1988) Control and Optimization in Catalytic Reforming, M.A.Sc. Thesis, University of Waterloo, Waterloo, Ontario, Canada.

Isidori A. (1989) Nonlinear Control Systems: An Introduction. 2nd ed Springer Verlag, New York.

Labossiere G.A. (1990) Modelling and Control of Blast Furnace Stoves. M.Eng.Sci Thesis, Department of Chemical Engineering, The University of Queensland, Queensland 4072, Australia.

Lee P.L. (1991) Direct Use of Nonlinear Models, Proc. CPC IV, South Padre Island, Texas, USA 517-542.

Lee P.L. and Sullivan G.R. (1988) Generic Model Control (GMC). Comput. & Chem. Eng 12: 573-580

Lee P.L., Newell R.B. and Sullivan G.R. (1989) Generic Model Control - A Case Study. Can J Ch E, 67:478-484

Lee P.L., Zhou W., Cameron I.T., Newell R.B. and Sullivan G.R. (1991) Constrained Generic Model Control of a Surge Tank. Computers & Chem Eng.15,3:191-195

Lee P.L., Zhou W., Newell R.B. and Greenfield P.F. (1988) Generic Model Control of an activated sludge plant. First Regional Seminar on Process Control, Kuala Lumpur, Malaysia 28-29 November 222-235.

Liu J., Pieper J. and Surgenor B. (1990) Intelligent identification and control of a thermal/hydraulic process: SISO PSTC and MIMO GMC. Internal Report. Manuf. Research Corp. of Ontario, Canada

Malik A. (1988) Dual Composition Control of an Industrial Propylene Splitter. IFAC Symposium on Model Based Process Control, Atlanta GA, USA

MATLAB (1990) Users Guide. The MathWorks Inc. Natick, MA USA

Newell R.B. and Lee P.L. (1989) Applied Process Control - A Case Study. Prentice Hall, Sydney, Australia

Riggs J.B. (1989a) Nonlinear Process Model-Based Control of a Propylene Sidestream Draw Column. Ind Eng Chem Res. 29,11:2221-2226

Signal P.D. (1992) GMC Relevant Modelling: Tools for an Improved Modelling Methodology, PhD Thesis, The University of Queensland, Queensland 4072, Australia.

Signal P.D. and Lee P.L. (1992) Generic Model Adaptive Control, Chem Eng Commun. 115:35-52

Sinha R. and Riggs J.. (1989)   High Purity Distillation Control Using Nonlinear Process Model-Base Control. Advances in Instrumentation, Proc ISA/89 Paper #89-0531, Vol 44, Part 2, 765-772

Stephanopoulos G. (1984) Chemical Process Control - An Introduction to Theory and Practice. Prentice Hall, Inc., Englewood Cliffs, New Jersey.

Surgenor B.W. and Hesketh T. (1989) Multivariable control of a furnace: Optimal LQS versus model based GMC. American Control Conf. June 21-23, Pittsburgh, PA, USA.

Surgenor B.W. and Wijesuriya E.T. (1990)   Control of a Pneumatic Positioner: Simulation results for Model-Based versus Conventional Control.   IEEE Indust Automation Conference, June 19-21, Toronto, Canada

Wilson D.I. (1990) Advanced control of a batch raw sugar crystallizer. PhD Thesis, Department of Chemical Engineering, The University of Queensland, Queensland 4072, Australia.

Wilson D.I., Lee P.L., White E.T. and Newell R.B. (1988) Computer control of an industrial sugar crystallizer. Process Systems Engineering Conf, Sydney, Australia 28 August - 2 September 26-31.

Zhou W. and Lee P.L. (1990) Model-based controller design for a heavy oil fractionator. Control 90, 4th Australian Control Conference, 1-3 August 12-16.

Zhou W., McNamara A.R., Lee P.L., Clark C.C., Lock Lee L. and Burgess J.M. (1992a) Simulation Studies of Model-Based Controller Design for an Ironmaking Blast Furnace. CHEMECA 92, Canberra Australia, September 27-30 199.1-205.1.

Zhou W., Lee P.L. and Sullivan G.R. (1992b) Robust Stability Analysis of Generic Model Control. Chem Eng Commun. 17:41-72

CHAPTER 3

AN INDUSTRIAL APPLICATION OF REFERENCE SYSTEM SYNTHESIS/GENERIC MODEL CONTROL: WASTEWATER pH CONTROL

3.1 ABSTRACT

A nonlinear controller designed by reference system synthesis was applied to control the pH of wastewater being discharged from a petrochemical plant. A first principles model was developed based on a set of simplifying assumptions, and validated against plant data. The controller was implemented as a Fortran program in a PMX computer linked to TDC2000. The results were the essentially complete elimination of the need for manual intervention when major disturbances arose, and significantly improved control within the regulatory consent limits for effluent pH.

3.2 INTRODUCTION

The plant was experiencing chronic problems controlling its effluent pH. The number of incidents where the effluent exceeded the regulatory consent limits was excessive. Neutralizing agent consumption was higher than necessary because of controller cycling. This situation existed regardless of a well-designed effluent treatment system. Tuning the pH controllers was a rite of passage for application engineers assigned to the plant. Despite the efforts of many good people, no set of controller settings provided satisfactory performance over all conditions. Instead of trying to tune again, we decided to implement a nonlinear controller designed by reference system synthesis (Bartusiak, et al., 1989).

Controller design by reference system synthesis is a three step process. First, a dynamic model of the process must be developed. Second, the desired closed loop behavior of the plant is specified in the form of a differential or integro-differential equation (*i.e.* the reference system). Third, the control law is derived by minimizing the difference between the process model and the reference system. In simple cases, this minimization can be done analytically which results in a closed form solution for a nonlinear control law. The Reference System Synthesis method is equivalent to the Generic Model Control (GMC) method (Lee and Sullivan, 1988).

Reference System Synthesis and Generic Model Control are but two techniques in an active research area in chemical engineering. Bequette (1991) has published a review which sets reference system synthesis and generic model control within the context of this research.

The literature on pH control is itself extensive. A rough categorization of the methods applied for pH control is the following: non-model based, nonlinear model based, and adaptive techniques. Shinskey blazed many of the trails in the pH control field, and his reference (1988) discusses the non-model based techniques -- PID controllers with nonconstant gains, such as error-squared and three-piece nonlinear. In the same reference, he also discusses the process design and instrumentation matters for pH control which will be most helpful to designers and control engineers in industry.

The controller described in this chapter is an instance of the nonlinear model based type. Other examples of nonlinear model based pH controllers are described by Wright and Kravaris (1991), Parrish and Brosilow (1988), and Williams, et al. (1990). Other nonlinear model based techniques and the adaptive control approaches for pH control are surveyed in the review by Gustafsson and Waller (1992).

In this chapter, we provide a process description, and detail the development and validation of a simple first principles model for pH dynamics. Next, we discuss the reference system synthesis controller derivation, tuning, and implementation. Finally, we present plant data which illustrate the nonlinear controller's performance.

3.3 PROCESS DESCRIPTION

The plant produces approximately 210 metric tonnes per hour (950 gpm) of wastewater during normal operating conditions. Several of the most noxious streams entering the effluent system are treated at-source. However, all effluent from the plant is routed together for final treatment. The final treatment process take place in a concrete-lined pit in which oil separation, primary pH adjustment and secondary pH adjustment are applied.

A schematic of the effluent treatment pit is provided in Figure 3.1. In the figure, flow is from right to left. Rightmost in the schematic are the oil separators. The "Coarse Pit" designator refers to the first stage of pH adjustment in which either 96% sulfuric acid or 20% caustic solutions are added via a compound split range valve arrangement (four valves). After the Coarse Pit, an equalization basin is provided to dampen major pH swings. Final pH adjustment is applied in the "Fine Pit" where, again, either 96% sulfuric acid or 20% caustic solutions are added via a split range valve setup (two valves). At the extreme left of Figure 3.1 is the pump bay from which the effluent can be either discharged to the environment, or recycled to a basin adjacent to the effluent treatment pit if the wastewater is outside of the consent limits for pH or Total Oxygen Demand.

Duplicate pH meters are provided at the locations indicated in the Coarse and Fine Pits. A rough estimate of effluent flow rate is measured using a V-notch weir just upstream of the Coarse Pit. Temperature is measured in the pipe downstream of the discharge pumps. The neutralizing agents are fed by gravity from elevated storage tanks through the control valves. There are no flow measurements on the acid and caustic streams.

The nonlinear controller discussed in this chapter was implemented for the Fine Pit.

3.4 MODEL DEVELOPMENT

From a study of the various streams entering the effluent system, we concluded that the predominant acid/base species in the wastewater were derived from sulfuric acid and caustic soda, and that there was little buffering

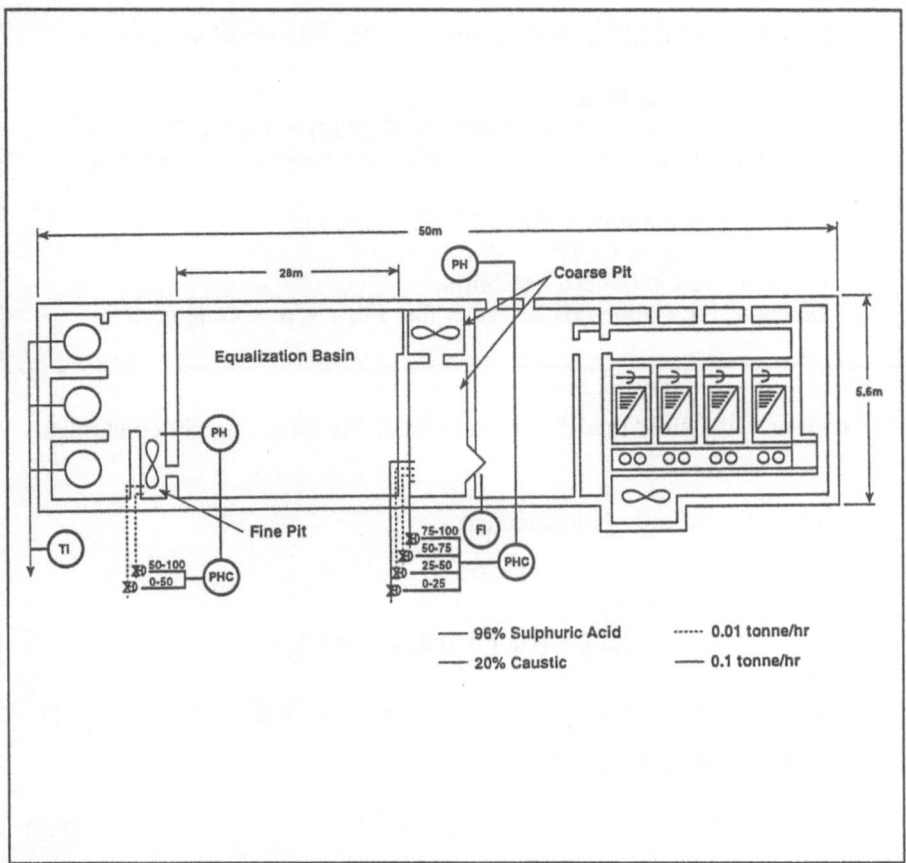

Figure 3.1 Schematic of the effluent treatment unit.

potential.   These and other simplifying assumptions, summarized in Table 3.1, were made as the basis for modelling.

Table 3.1 Modelling Assumptions

Modelling Assumptions
1. Sulfuric acid (H_2SO_4) and caustic (NaOH) only acid/base species present. 2. Insignificant buffering. 3. Complete acid/base dissociation. ($K_a(HSO_4^-) = 1.2 \times 10^{-2}$ but at the near-neutral conditions of the setpoint, dissociation is practically complete.) 4. Fine Pit is a constant volume CSTR. 5. Perfect mixing in CSTR. 6. Infinitely fast acid/base equilibria. 7. Linear valve characteristics, such that Flow = $\alpha \times$ Output.

The nonlinear dynamic model is derived from the following six conditions.

• Electrical neutrality is maintained.

$$[H^+] + [Na^+] = [OH^-] + 2[SO_4^=] \tag{3.1}$$

• The water dissociation equilibrium fixes the relative concentration of hydrogen and hydroxyl ions.

$$[H^+][OH^-] = K_w(T) \tag{3.2}$$

- The fluxes of sodium and sulfate ions through the Fine Pit are described by material balance.

$$\frac{d}{dt}[Na^+] = \left(\frac{F_i}{V}\right)[Na^+]_i + \left(\frac{F_c}{V}\right)[Na^+]_c - \left(\frac{F_{out}}{V}\right)[Na^+] \tag{3.3}$$

$$\frac{d}{dt}[SO_4^-] = \left(\frac{F_i}{V}\right)[SO_4^-]_i + \left(\frac{F_a}{V}\right)[SO_4^-]_a - \left(\frac{F_{out}}{V}\right)[SO_4^-] \tag{3.4}$$

- The pH definition relates the instrument reading to the hydrogen ion concentration.

$$pH = -\log[H^+] \tag{3.5}$$

- The effluent material balance, assumption 4. in Table 3.1, and an additional assumption that the neutralizing agent flows are negligible relative to the effluent flow yield the sixth basis condition.

$$\frac{dV}{dt} = 0 = F_i + F_a + F_c - F_{out} \tag{3.6}$$

$$\left(F_i \gg F_a, F_c\right) \rightarrow F_i \approx F_{out} \equiv F$$

To proceed with the model derivation, we first substitute the water equilibrium relation 3.2 into the condition of electoneutrality.

$$[H^+] - \frac{K_w}{[H^+]} = 2[SO_4^-] - [Na^+] \tag{3.7}$$

Differentiation of equation 3.7 with respect to time yields the hydrogen ion

49

dynamics in terms of the sodium and sulfate dynamics.

$$\left(\frac{d}{dt}[H^+]\right)\left(1 - \frac{K_w}{[H^+]^2}\right) = 2\frac{d}{dt}[SO_4^=] - \frac{d}{dt}[Na^+] \tag{3.8}$$

We differentiate the pH definition to relate the hydrogen ion dynamics to pH dynamics.

$$\frac{d}{dt}[H^+] = \frac{d}{dt}\left(10^{-pH}\right) = \left(-2.303 \times 10^{-pH}\right)\frac{d}{dt}(pH) \tag{3.9}$$

Substituting the sodium and sulfate ion material balances 3.3 and 3.4 together with equation 3.9 into equation 3.8, we derive the pH dynamics in terms of flow rates and ionic concentrations.

$$\frac{d}{dt}(pH) = \left(\frac{-1}{2.303(10^{-pH} + K_w 10^{pH})V}\right)\left\{2\left(F[SO_4^=]_i + F_a[SO_4^=]_a - F[SO_4^=]\right)\right.$$
$$\left. -\left(F[Na^+]_i + F_c[Na^+]_c - F[Na^+]\right)\right\} \tag{3.10}$$

Since the effluent entering and leaving the Fine Pit must also satisfy the electroneutrality condition, we can relate the ionic concentrations as described in equations 3.11a and 3.11b.

$$[SO_4^=]_i = \frac{1}{2}\left([Na^+]_i + 10^{-pH_i} - K_w 10^{pH_i}\right) \tag{3.11a}$$

$$[SO_4^=] = \frac{1}{2}\left([Na^+] + 10^{-pH} - K_w 10^{pH}\right) \tag{3.11b}$$

Finally, by substituting equations 3.11a and 3.11b into 3.10 and simplifying, we derive an expression for the pH dynamics in terms of values which are known (K_w, V, $[SO_4^-]_a$, $[Na^+]_c$), measured (pH, F, T), or estimatable (pH_i, F_a, F_c).

$$\frac{d}{dt}(pH) = \left(\frac{-1}{2.303(10^{-pH} + K_w 10^{pH})V} \right) \{ F(10^{-pH_i} - K_w 10^{pH_i}) + 2F_a[SO_4^-]_a - $$
$$F(10^{-pH} - K_w 10^{pH}) - F_c[Na^+]_c \}$$

(3.12)

3.5 MODEL VALIDATION

To validate the model, we first developed estimates for pH_i, F_a and F_c from the available measurements. The pH of the effluent entering the Fine Pit was estimated by averaging five values of the Coarse Pit pH measurement delayed over a window of time centered about 19 minutes (*i.e.* 15, 17, 19, 21 and 23). We arrived at this simple estimation algorithm via engineering judgement by (1) matching the deadtime across the Equalization Basin and (2) smoothing to model the convective/diffusive processes.

The acid and caustic addition rates were inferred from their valve outputs with the simplifying assumption that the installed characteristics (Luyben, 1990) of the valves were linear equation 3.13. Initial estimates of the $F_{j,ident}$ factors were obtained from a few steady state data points. The factors were fine-tuned while the model was running on-line to fit equation 3.12 to the

51

plant data.

$$F_j = F_{j,ident} \times \left(\frac{M_{j,max}}{\rho_j} \right) \times \left(\frac{OP_j - OP_{j,min}}{OP_{j,max} - OP_{j,min}} \right) \qquad (3.13)$$

Nominal values for the data of interest in equations 3.12 and equation 3.13 are listed in Table 3.2.

Table 3.2 Nominal values of process data

F	53.0 l/sec	$M_{a,max}/\rho_a$	0.00281 l/sec
V	17600 l	$M_{c,max}/\rho_c$	0.00278 l/sec
$[SO_4^-]_a$	18.0 moles/l	T	25.0 °C
$[Na^+]_c$	6.09 moles/l		

One example of the model fit is illustrated in Figure 3.2. The model tracked the process quite well, and we proceeded to implement the controller.

3.6 CONTROLLER DESIGN AND TUNING

To design a controller by the Reference System Synthesis/Generic Model Control method, we take (1) the process model, and (2) a specification for the desired closed loop behavior of the process (*i.e.* the reference system), and then (3) compute the control law by minimizing the difference between the open loop and reference systems. In our case, the process model is given by equation 3.12.

The "no frills" specification for the desired closed loop behavior is the integro-differential equation 3.14 which yields proportional and integral action in the controller. The coefficients k_1 and k_2 are the tuning parameters which specify the closed loop speed of response.

$$\frac{d}{dt}(pH)^* = k_1 \left(pH^{set} - pH\right) + k_2 \int_0^t \left(pH^{set} - pH\right) dt \qquad (3.14)$$

The control law is derived by minimizing the difference between the model and the reference system (i.e. $d(pH)/dt - (d(pH)/dt)^* = 0$).

$$2\frac{F_a}{V}[SO_4^-]_a - \frac{F_c}{V}[Na^+]_c = \frac{F}{V}\left\{(10^{-pH} - K_w 10^{pH}) - (10^{pH_i} - K_w 10^{pH_i})\right\}$$

$$+ -2.303(10^{-pH} + K_w 10^{pH})\left\{k_1(pH^{set} - pH) + k_2 \int_0^t (pH^{set} - pH) dt\right\}$$

$$(3.15)$$

Note that both acid and caustic flow rates appear on the left hand side of equation 3.15. These flows are mutually exclusive. Hence, in the controller, we compute equation 3.15 twice -- once for F_a and once for F_c -- and only implement the nonnegative value.

Initial values for the tuning constants k_1 and k_2 were determined by pole placement. The specification for the closed loop response was that the pH dynamics should be roughly twice as fast as the residence time of the Fine Pit, and that the response exhibit critical damping. Consider the closed loop characteristic equation derived from the reference system equation 3.14.

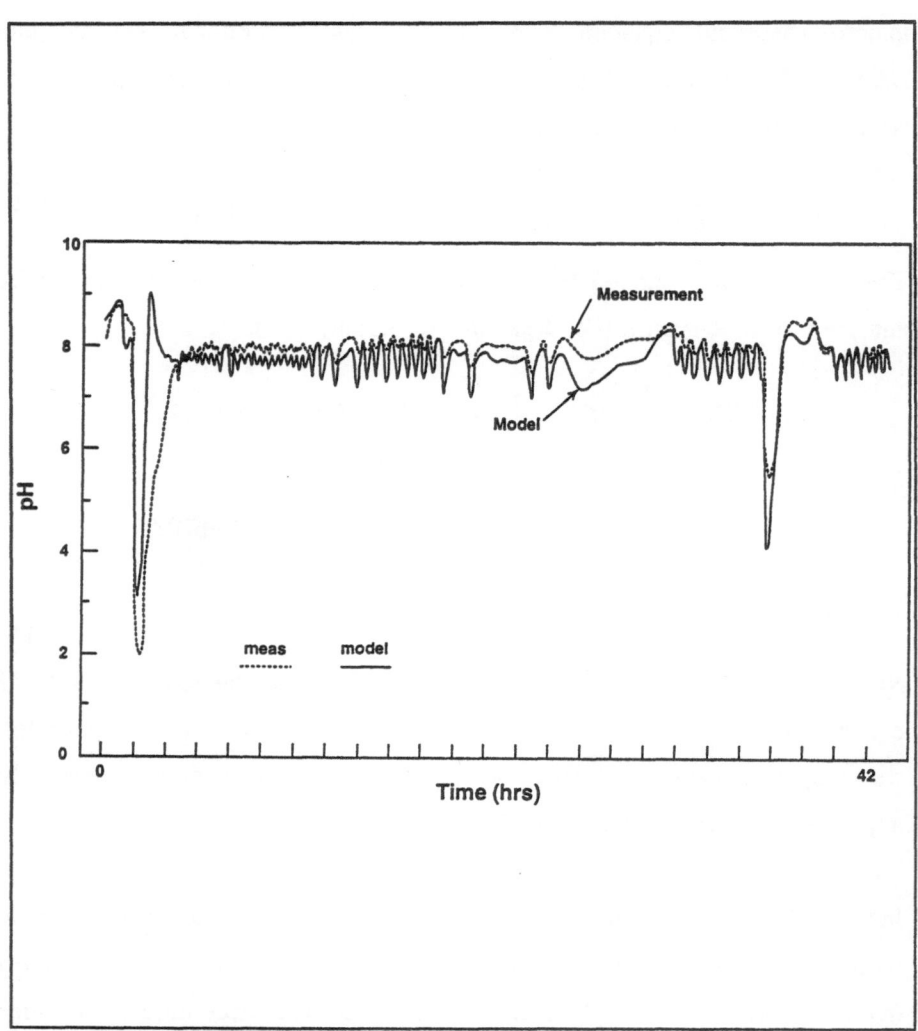

Figure 3.2 pH model fit to plant data

$$s^2 + k_1 s + k_2 = 0 \qquad (3.16)$$

The verbal tuning specification translates to repeated roots of 3.16 at a target pole twice the distance from the origin in the left half plane as the open loop pole associated with the residence time.

$$k_1^2 - 4k_2 = 0$$

$$\frac{-k_1}{2} = \lambda_{target} \qquad (3.17)$$

The residence time was approximately 8 minutes. Therefore, the target pole was -0.25 min^{-1}, and the initial values of k_1 and k_2 were 0.5 min^{-1} and 0.0625 min^{-2} respectively.

3.7 IMPLEMENTATION

The controller was implemented in a PMX process control computer above TDC2000. Initially we coded the model in a BPL program (A BPL is a procedural program which has the advantage of tight integration with the process control computer but also has several disadvantages such as only single precision arithmetic). During model validation, however, we became concerned about the limited numerical precision by 24 bit word lengths. The final phases of the modelling and the control algorithm were programmed in PMX FORTRAN in which double precision floating point arithmetic was used. The controller runs every fifteen seconds.

The controller was implemented as an application above the existing linear controller. Accordingly, additional code was written to convert the results of the control law equation 3.15 into the output of the split range controller above the control valves. To enable fine-tuning of the model online, we provided fudge factors $F_{j,fudge}$ which, in essence, adjusted the identification factors $F_{j,ident}$ and hence the process gains in the model. These fudge factors allowed us to compensate for the uncertainty in the actual flowrates of the neutralizing agents relative to the valve manufacturing's maximum raings.

$$OP_j = OP_{j,min} + \left(\frac{F_j \rho_j}{F_{j,fudge} F_{j,ident} M_{j,max}} \right)(OP_{j,max} - OP_{j,min}) \qquad (3.18)$$

The control application in PMX was built in two levels - feedback-only and feedforward/feedback. The feedback-only control law was equation 3.15 with nominal constant values for pH_i, F and T. This provided the operators with the capability to switch levels which allowed the controller to degrade gracefully in th event of the loss of the instrument for pH_i, F or T. In fact, some time after the controller was commissioned, the Equalization Basin was cleaned out after a turnaround. The cleanout changed the fluid flow dynamics significantly which invalidated the pH_i estimate described in the Modelling section (*i.e.* greater attenuation of pH disturbances was achieved). The operators were satisfied running in feedback-only mode, so we have no incentive to revisit the pH_i estimate, or to request that an additional pH instrument be added at the inlet to the Fine Pit.

Regarding tuning, we modified the k_1 and k_2 tuning parameters away from their initial theoretical values during commissioning. The motivation was to increase the proportional action and reduce overshoot. The final values of k_1 and k_2 were 0.67 min^{-1} and 0.02 min^{-2} respectively.

Finally, several other practical implementation matters deserve brief comments.

- Reset windup was mitigated by restricting the absolute value of the integral error to a user-set maximum value.
- The incremental change in the control law result could be constrained to a user-specified step bound.
- The controller was coded with a provision for a deadband around the pH setpoint. However, there was no need to use this feature based on the performance we achieved.

3.8 RESULTS

It is instructive to consider the performance of the linear controller before discussing the nonlinear controller. The general strategy was to control aggressively in the Coarse Pit and gently in the Fine Pit. This strategy was manifested both in the process design for the neutralization system, and in the tuning constants of the linear controllers (which were continually being readjusted by the application engineers). In practice, the Coarse Pit was often in a limit cycle racing back and forth across the neutral zone after a load disturbance occurred, while little automatic control was being applied in the Fine Pit. Furthermore, the Fine Pit controller would often windup (even with anti-windup measures) and, for example, continue to add acid when the pH was consistently low. The operators had to routinely put the Fine Pit controller into manual and reset the output to 50% (both acid and caustic valves closed) to get the controller back "in the ballpark".

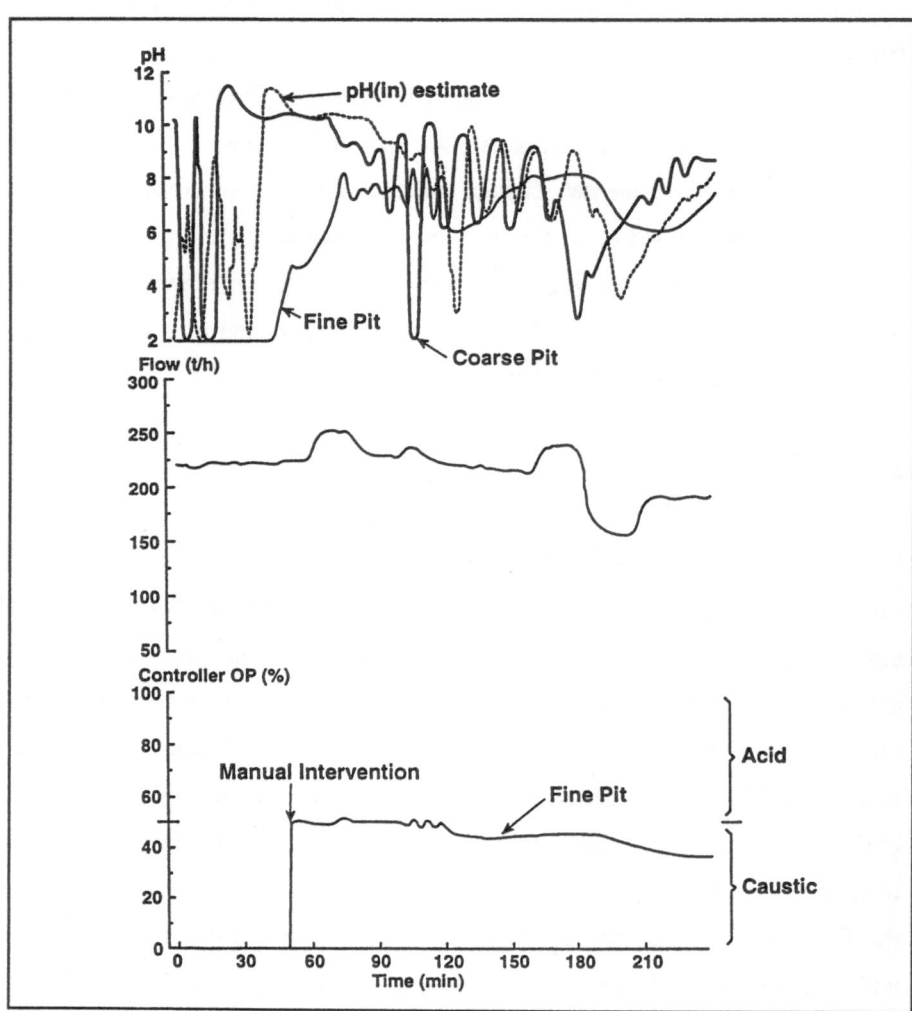

Figure 3.3 Typical response of linear controller

These phenomena are illustrated in Figure 3.3. At the beginning of the plots, the Coarse Pit is cycling; the Fine Pit effluent is acidic and being recycled within the plant. At time 25, the Coarse Pit stops cycling with the effluent in the alkaline range. Seeing the Fine Pit controller wound up, the operator put the loop into manual at time 50; reset the output to 50%; and returned the loop back to automatic. Little regulation is achieved by the Fine Pit controller during the period from 50 to 240 minutes -- the controller output does not move much and the pH of the effluent leaving the Fine Pit generally follows the pH_i estimate.

Now let's consider the performance of the nonlinear controller. Figure 3.4 depicts a characteristic load disturbance, *i.e.* a sharp acid spike. The nonlinear controller acts vigorously when the pH is far from the neutral range, and gently when the pH is near-neutral. During the time period 70 to 210, the controller strokes the caustic valve and then the acid valve fully open as the acid spike comes followed by a surge of alkaline effluent. As the pH returns to setpoint after time 240, the controller moves are small in the region where the break in the titration curve occurs. Also note that the nonlinear controller compensates for the difference in the ionic strength of the neutralizing agents as evidenced by the sudden change of slope of the output at time 370.

Despite the vigorous action of the nonlinear controller, it was not prone to high frequency cycling across the neutral zone. In Figure 3.5, the Coarse Pit is cycling symmetrically around a pH of 7. The Fine Pit controller continues to respond actively to the cycling pH_i estimate, but does not begin to cycle itself. Figures 3.4 and 3.5 both illustrate the controller performance with feedforward action enabled.

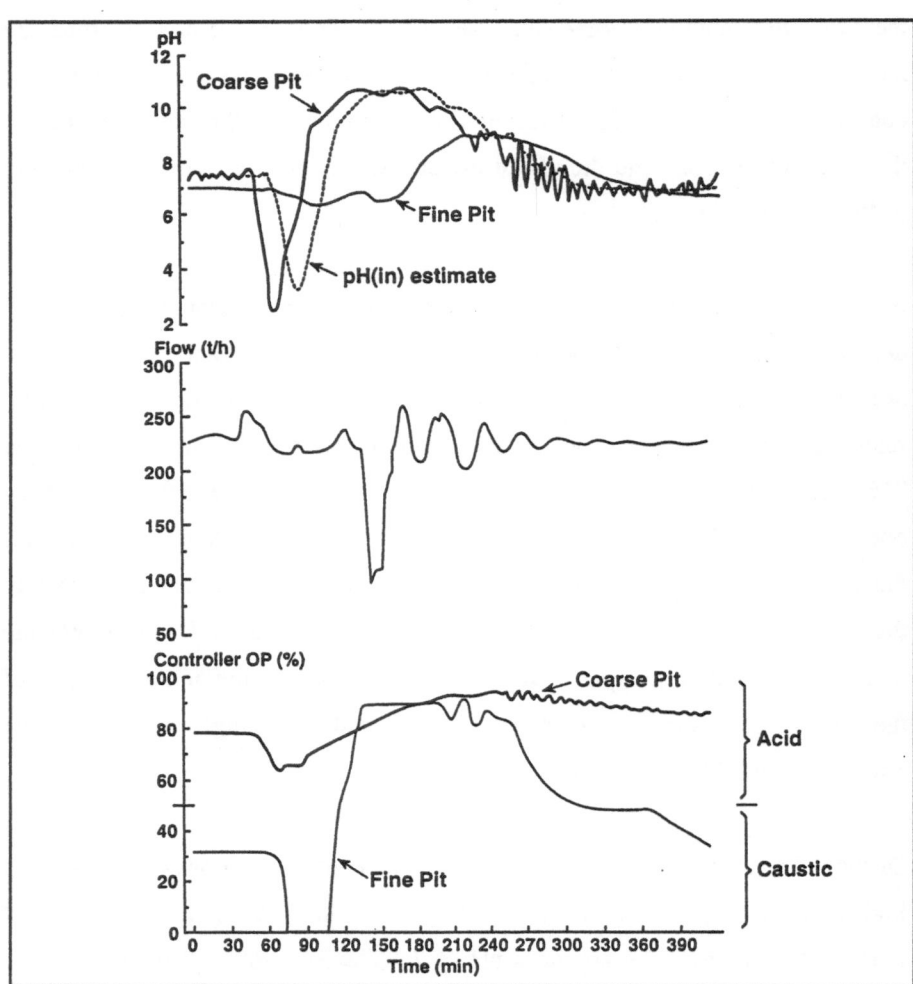

Figure 3.4 Nonlinear controller response to acid spike disturbance

Although the nonlinear controller's performance was significantly better than the linear controller, it did not achieve the desired linearization as expressed in the reference system equation 3.14. The speed of response when returning from the alkaline region (*e.g.* from pH 8.5 to the setpoint at 7.6) was faster than a return from the acidic region (*e.g.* from 6.5 to 7.6). Furthermore, during times of no load disturbances, the nonlinear controller did exhibit a low frequency limit cycle (period on the order of an hour) of modest amplitude (pH between 7 and 8 for a setpoint at 7.6). However, from the perspective of the operators and for environmental compliance, this limit cycle was inconsequential.

3.9 CONCLUSIONS

The reference system synthesis/generic model control methodology was effective for designing and implementing an industrial-quality nonlinear controller for regulating wastewater pH. Compared to the existing linear controller, the nonlinear controller provided aggressive regulation without high-frequency cycling through the neutral zone, and essentially eliminated the need for manual intervention.

The source of most of the problems during model validation and controller commissioning was the lack of knowledge of the actual flow rates of the neutralizing agents.

In this application, a simple strong acid/strong base model was sufficient. Equation 3.12 is probably the least complicated model for pH dynamics one

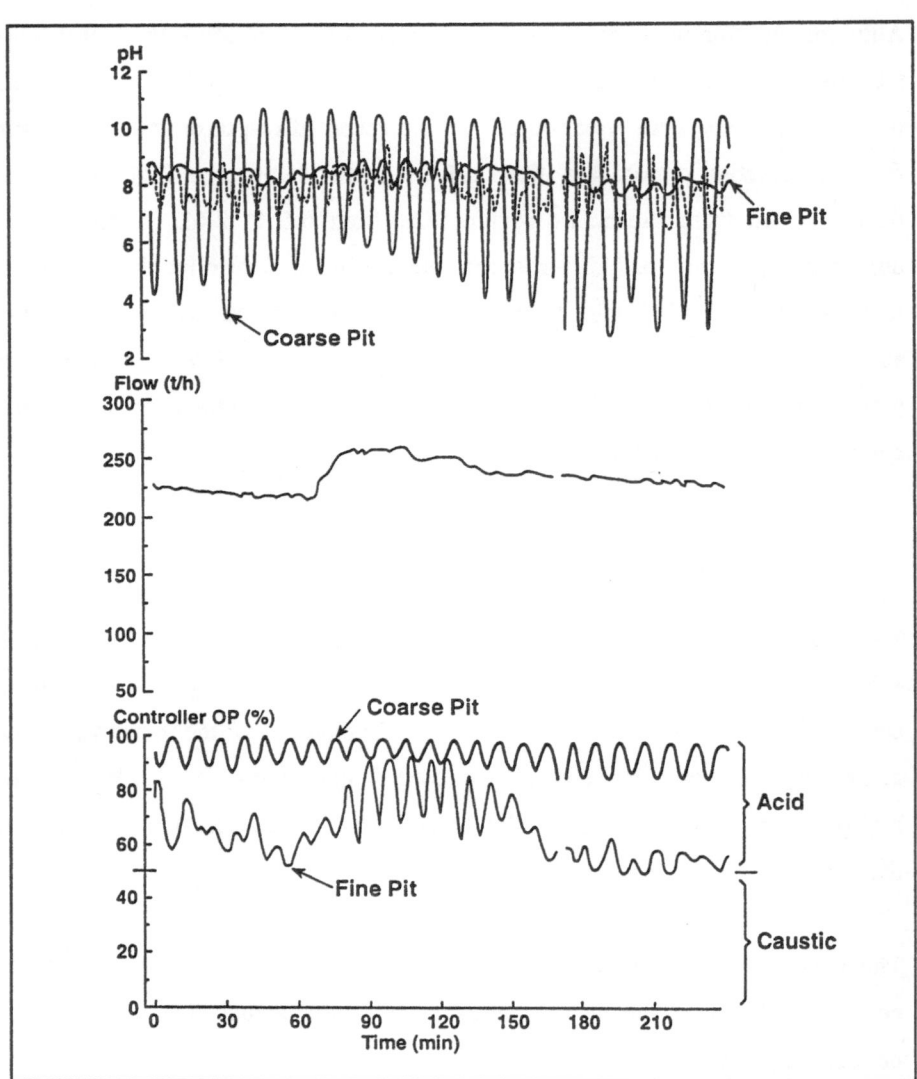

Figure 3.5 Nonlinear controller response to cycling in Coarse Pit

could derive. A rigorous model for systems of weak acids and bases with buffering species will include terms for ionic concentrations whose values are

62

not readily measurable in industrial situations (Gustafsson and Waller, 1992).

Could we have achieved satisfactory control with non-model based nonlinear techniques, such as error squared, three-piece nonlinear, or scheduled gains (Shinskey, 1988)? I have no doubt that these techniques could have made significant improvements in performance, as Shinskey has shown, with less engineering time invested. However, the model based controller provides unique capabilities for controlling to pH setpoints other than 7, and for compensating for the differing ionic strengths of the neutralizing agents. The majority of the engineering time was spent on developing and validating the pH model, not designing and implementing the controller. Once the modelling methodology was specified, the engineering work proceeded in a straightforward manner.

3.10 ACKNOWLEDGEMENT

This work was accomplished with the encouragement and assistance of Ken Walker, Mark Harrington, Martin Kerlin and Ronnie Millet.

3.11 NOMENCLATURE

F   volumetric flow rate (l/sec)

$K_w(T)$ water dissociation constant ($moles^2/l^2$)

k_1, k_2 tuning constants in reference system (sec^{-1} and sec^{-2}, respectively)

M   mass flow rate (g/sec)

OP   output to valve (%)

s   the Laplace operator

t   time (*sec*)

T   temperature (°*C*)

V   volume (*l*)

[ ]   concentration (*moles/l*)

Greek symbols

α   general constant

ρ   density (*g/l*)

λ   pole in imaginary plane (time^{-1})

Subscripts

a   sulfuric acid stream

c   caustic stream

fudge online correction factor

i   effluent stream into Fine Pit

ident model identification factor

j   general stream index; either a for acid or c for caustic

out   effluent stream out of Fine Pit

3.12 REFERENCES

Bartusiak R.D., Georgakis C. and Reilly M.J. (1989)   Nonlinear feedforward/feedback control structures designed by reference system synthesis. Chem. Eng. Sci. 44(9):1837-1851.

Bequette B.W. (1991) Nonlinear process control: A review. Ind. Eng. Chem. Res. 30: 1391-1413

Gustafsson T.K. and Waller K.V. (1992) Nonlinear and adaptive control of pH. Ind. Eng. Chem. Res. 31:2681-2693.

Lee P.L. and Sullivan G.R. (1988) Generic Model Control (GMC). Comput Chem. Eng. 12:573-580.

Luyben W.L. (1990) Process Modelling, Simulation and Control for Chemical Engineers. McGraw-Hill, New York.

Parrish J.R. and Brosilow C.B. (1988) Nonlinear inferential control. AIChEJ. 34 (4):633-644.

Shinskey F.G. (1988) Process Control Systems: Application, Design and Tuning. McGraw-Hill, New York.

Williams G.L., Rhinehart R.R. and Riggs J.B. (1990) In-line process-model-based control of wastewater pH using dual base injection. Ind. Eng. Chem. Res. 29:1254-1259.

Wright R.A. and Kravaris C. (1991) Nonlinear control of pH processes using the strong acid equivalent. Ind. Eng. Chem. Res. 30: 1561-1572.

CHAPTER 4

USING TRAY-TO-TRAY MODELS FOR DISTILLATION CONTROL

4.1 INTRODUCTION

It has been estimated that there are more than 40,000 distillation columns in operation in the refining and petrochemical industries in the U.S. (Humphrey et. al., 1991) which accounts for 95% of all separation process in these industries. It is further estimated that distillation consumes approximately 3% of the total U.S. energy consumption.   More importantly, distillation operations almost exclusively are responsible for the purity and uniformity of the products for the petrochemical industry.

As a result, distillation control has a significant economic impact on the refining and chemical industries.   Therefore, improved distillation control across the board would result in major economic improvement. For example, it has been estimated (Humphrey et. al, 1991) that there is potential for an average 15% reduction in the energy consumption by distillation in the refining and chemical industries in the U.S. with improved control. Specifically, one or more of the following economic benefits can result from improved distillation control:

- Reduced product variability
- Reduction in the rate of production of off-specification products
- Increased production rates when a column is the bottleneck in the system
- Reduced utility usage
- Increase in the yield of a product through an increase in the average impurity levels while maintaining product specification.

In the refining industry, the most important benefit is the reduced utility usage, but when a column represents a bottleneck, larger economic savings are possible.

For the chemical industry, the most important benefit is reduced product variability. With the advent of product certification procedures (e.g., the ISO 9000 series), a greater emphasis is being placed upon the reduction of product variabilities. In fact, the consumers of chemical intermediates are beginning to request tighter and tighter specifications for the uniformity of their feedstocks. For example, polymer manufacturers are more concerned with receiving a uniform feedstock than the purity of the feedstock. As a result, suppliers which may provide on the average a higher purity feedstock are losing business to suppliers that provide the feedstock with significantly reduced variability even with higher average impurity levels. This is true because the polymer manufacturers makes a better product when their feedstocks are uniform. This phenomena is spreading as manufacturers recognize the benefits of reduced variability feedstocks. Distillation is a challenging process control problem due to the inherent nonlinearity of the process and the severe coupling. Moreover, columns are many times subjected to large upsets in feed composition and feed flow rate. In addition, nonstationary behaviour, such as a variation in overall tray efficiency, can result.

A typical distillation column is shown in Figure 4.1. In a typical column, the liquid level in the reflux drum and the pressure are controlled by simple PI controllers. The variables that most influence economic performance are the compositions of the two product streams, one at the top of the column and one at the bottom of the column. These can be controlled by adjusting differing combinations of two flowrates, but include reflux ratio, reflux rate, distillate rate, reboil rate, bottoms product rate.

Table 4.1 summarizes the advantages and disadvantages of PID, model predictive control (MPC) (Cutler and Ramaker, 1979), artificial neural nets (Bhat and McAvoy, 1990), and nonlinear process model based control (nonlinear PMBC) (also known as GMC, Lee and Sullivan, 1988). Note that while PID and MPC are generic, they both assume a linear picture of the process. Since nonlinear PMBC uses a nonlinear process model directly for control, it is a nonlinear controller. Artificial neural nets use empirical nonlinear models.

Table 4.2 lists the base case conditions for a depropanizer with 39 theoretical stages producing a high purity overhead product and a moderate purity bottoms product. Table 4.3 lists the process gains for a reboil-reflux (V-L_o)

configuration calculated using a 50% relative change in impurity levels. Note the nonlinearity of the gains evident from the changing process gains for step

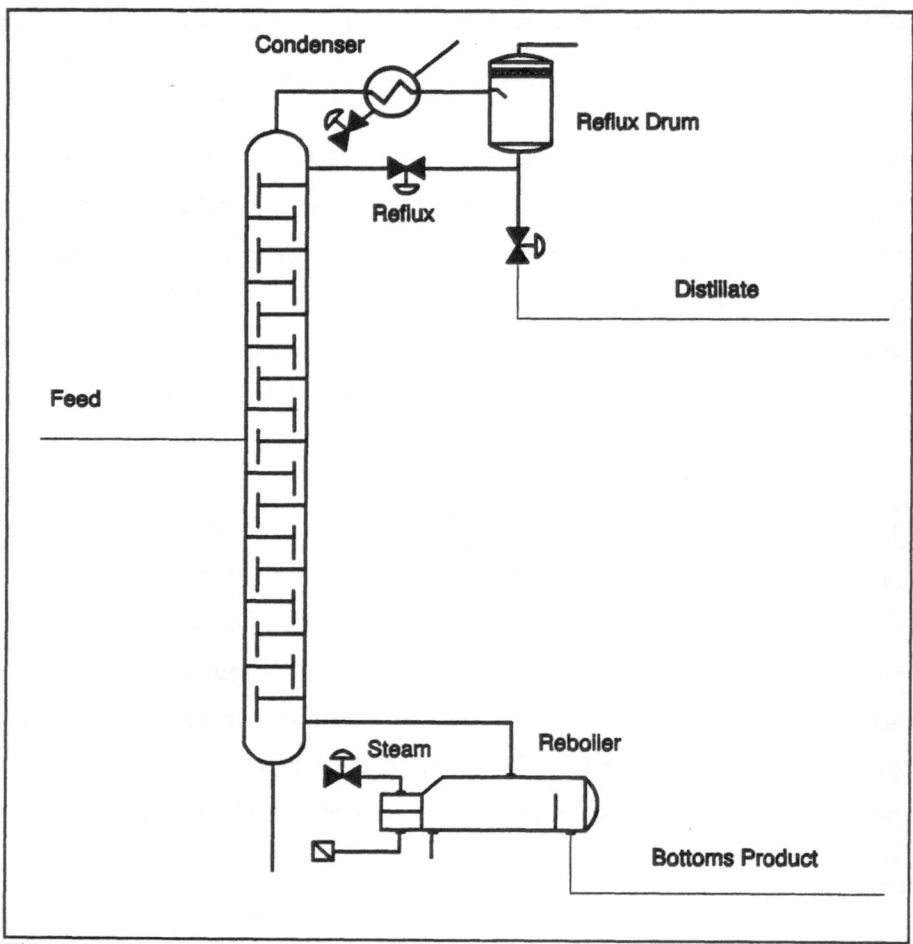

Figure 4.1 Typical Distillation Column

increases and step decreases of the manipulated variables. Also note that each manipulated variable affects each controlled variable, indicating a strong coupling between both inputs and both outputs.

Table 4.4 lists the average relative changes in the gains from the base case (Table 4.3) for a feed composition change, an operating pressure change, a feed rate change, a change in stage efficiency and a change in feed enthalpy. Since PID and MPC use fixed gain models, they are at an obvious disadvantage for the control of columns with significant variation in process gains. As a result, for such cases the PID and MPC controller must be de-tuned to maintain reliability at the expense of performance. The nonlinear PMBC controller has the potential of being able to predict these various gain changes if the process changes are measured and the model is sufficiently accurate.

4.2 DESCRIPTION OF COLUMNS CONSIDERED

Nonlinear PMBC work on several high reflux ratio, low relative volatility columns and a high relative volatility, multi-component column are described in this chapter. Following is an overview description of the columns considered in this work.

Low Relative Volatility Columns. At this point, we have applied nonlinear PMBC to four low relative volatility columns. While these columns were not strictly speaking purely binary columns, they are assumed to be binary columns for purposes of the controller model development. The sum of light key and heavy keys in the feed ranged from 80 to 90 mole percent. The

71

relative volatility between the light key and heavy key was approximately equal to 1.2 for each column. Each column had approximately 100 trays while the reflux ratios employed ranged from 10 to 50. As a result, the open loop time constants are estimated to range from 2 to 6 hours.

As with most distillation columns, the most difficult upset results from feed composition upsets. Only one column had an accurate on-line estimate of feed composition while all but one column had analyzers on both the overhead and bottom product streams. Composition analyzer delays ranged from 5 to 10 minutes. Since each of these columns produced a final product, statistical process quality (SPQ) charts were kept on them using the results from samples taken on each shift.

Table 4.1 Advantages and Disadvantages of Advanced Control Options

	Advantages	Disadvantages
PID	Availability Generic Well understood	Nonlinear process Strong coupling Nonstationary processes
MPC	Based upon empirical dynamic models Generic Provides decoupling Handles constraints directly	Uses linear models Assumes that process is stationary Significant identification process Requires large matrix operations

Neural Network	Uses nonlinear models Provides decoupling Generic	Needs to be trained for full ranges of operations Not well understood Training is CPU intensive
Nonlinear PMBC-GMC	Understands the nonlinearity Provides nonlinear decoupling Feedforward compensation Can adapt to nonstationary changes Tuned like a PI controller	Requires significant process knowledge Somewhat "tailor-made" Neglects dynamics except in tuning

Table 4.2 Base Case Values of Variables for the Depropanizer

Number of stages	=	39
Feed composition:		
Ethane	=	0.019
Propane	=	0.215
Isobutane	=	0.084
N-Butane	=	0.209
N-Pentane	=	0.159
N-Hexane	=	0.211
Feed rate	=	0.849 lbmoles/s
Feed temperature	=	165 degrees F
Feed tray location	=	18
Column pressure	=	18.040 atm
Stage efficiency	=	100 %
Composition of heavy key (isobutane) in top product	=	0.00364
Composition of light key (propane) in bottom product	=	0.02
Top product withdrawal rate	=	0.274 lbmoles/s
Reflux ratio	=	1.8
Temperature of column bottom	=	280 degrees F
Temperature of top tray	=	122 degrees F
Reflux temperature	=	104.9 degrees F

The flow controllers and the level controllers are pneumatic controllers. Composition controllers (whether PI or nonlinear PMBC) resided in a Perkin-Elmer control computer and performed as supervisory controllers since they selected the setpoints for the appropriate pneumatic flow controllers.

For three of the columns, the light and heavy keys were adjacent components on a relative volatility scale. But for one column, there is a component between the light and heavy key. As a result, for the latter case the use of a binary model was quite a bit more challenging.

Table 4.3 Base Case Gains

	$\Delta x/\Delta V$	$\Delta x/\Delta L$	$\Delta y/\Delta V$	$\Delta y/\Delta L$
increase in impurity level	-.295	-.529	-.302	-.193
decrease in impurity level	-.220	-.323	-.179	-.122

x - mole fraction impurity in the bottoms product
y - mole fraction impurity in the distillate product
V - vapor rate (# - moles/sec)
L_o - reflux rate (# - moles/sec)

Table 4.4 Relative Change in Process Gain From Base Case (Table 4.3)

Change	Δx/ΔV	Δx/ΔL	Δy/ΔV	Δy/ΔL
+5 Mole% change in Heavy Key	-14%	-30%	26%	25%
1.8 atm increase in column pressure	-10%	-15%	-16%	-15%
10% increase in Feedrate	10%	-11%	-11%	-10%
10% decrease in Stage Efficiency	-20%	-32%	-50%	-48%
20°F Increase in Feed Temperature	-5%	-10%	-13%	-11%

Finally, three of the columns have to operate against a condenser duty constraint at times during the summer months.

High Relative Volatility Multicomponent Columns. At this point, a nonlinear PMBC controller has been developed for dual composition control on a depropanizer and has been tested on a detailed dynamic simulator of the depropanizer. It remains to test the controller on the actual column.

Table 4.2 contains operating conditions that are typically observed. Note the multi-component character of the feed to the column. The column is equipped with online product analyzers for the overhead and bottoms products with cycle times of about 6 minutes but does not have a feed composition analyzer. The manipulated variables are the steam flow to the reboiler and the external reflux rate while the controlled variables are the propane concentration in the bottoms product and the iso-butane concentration in the distillate product. The accumulator level is controlled by the distillate flow rate and the reboiler level is controlled by the bottoms product flow rate.

At times during the summer, the column also encounters a condenser duty constraint.

4.3 CONTROLLER MODELS

Dynamic vs Steady State Models. With regard to the selection of a controller model for distillation column control, the first decision that must be made is whether to use dynamic or steady-state models. The application of GMC or other nonlinear PMBC controllers (e.g., nonlinear inferential model control, Economoric et.al., 1985; nonlinear predictive model control, Parrish and Brosilow, 1988) to distillation control using a dynamic column model requires (strictly speaking) measurements of the internal states (i.e., all tray compositions and liquid flow rates) which is not practical. Moreover, finding the model inverse of a dynamic distillation model is clearly computationally expensive. But certain assumptions can be made about the unmeasured states (e.g., initialized with steady-state results) allowing dynamic distillation models to be used in a nonlinear PMBC framework. Results reported by Pandit (1991) showed that while the dynamic model was computationally much more expensive than the steady-state model, the control results using both types of models were effectively equivalent.

GMC Control Law using Steady-State Models. Cott and Sullivan (1987)
showed that by assuming first order composition dynamics, the GMC control
law can be applied using steady-state model equations. For a distillation
column, target product composition levels, x_{ss} and y_{ss}, are calculated as
follows:

$$x_{ss} = x_o + K_{11}(x_{sp} - x_o) + K_{21}\int_0^{t_k}(x_{sp} - x)dt \qquad (4.1a)$$

$$y_{ss} = y_o + K_{12}(y_{sp} - y_o) + K_{22}\int_0^{t_k}(y_{sp} - y)dt \qquad (4.1b)$$

Then the values of y_{ss} and x_{ss} are used in the steady-state model to calculate
the control action. For example,

$$\begin{bmatrix} x_{ss} \\ y_{ss} \end{bmatrix} \rightarrow \begin{bmatrix} \text{Steady-State} \\ \text{Column Model} \end{bmatrix} \rightarrow \begin{bmatrix} V \\ L \end{bmatrix}$$

Figure 4.2 shows a schematic of the GMC control law. While the PI action
is not exactly a classical PI controller it is quite similar. If one considers the
steady-state model inverse with the process, the resulting system will be
linear on a steady-state basis if the model is perfect. Then the PI-type
controller will have a linear system to control which it might control quite
well. Therefore, in the limit of a perfect model this arrangement is expected
to work well as long as neither excessive measurements deadtime nor
excessive inverse action are present.

Note that when the proportional gain (dimensionless) terms (i.e., K_{11} and K_{12})

are equal to 1.0, the target values become the setpoint values plus any integral correction. Under these conditions, the controller is behaving as largely a feedforward controller with no proportional feedback.

Consider the case for which K_{11} is equal to 2, x_{sp} is equal to 1%, and x_o is 0.5%. Neglecting the integral term, x_{ss} is 1.5%. This means that if you want to move from too high a purity (0.5%) to your setpoint of 1%, you implement control that at steady-state (assuming a perfect model) would yield a value of x of 1.5%. Of course, we do not leave the process alone until steady-state. We re-evaluate the control action at each sample time. Since our control intervals are generally quite small compared to the time constant of an industrial column, the procedure is effective.

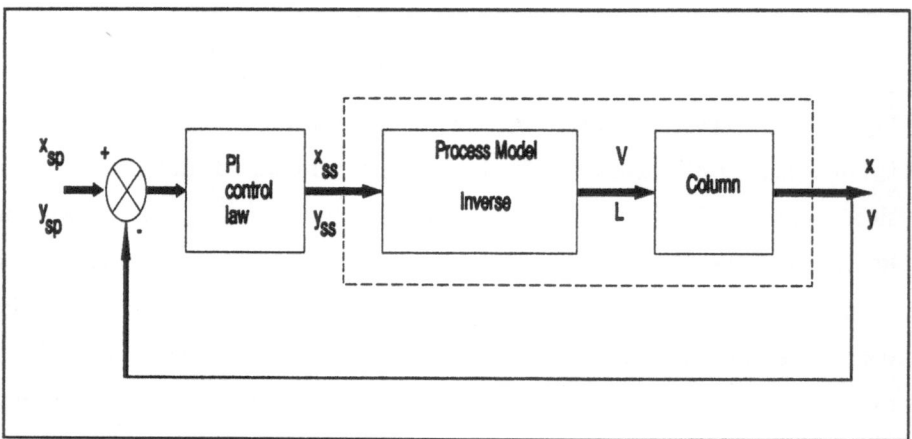

Figure 4.2 Schematic of the GMC control law

In effect, we are saying that the gain prediction and steady-state decoupling of the steady-state model is more important in distillation control than the dynamic behavior when the control interval is small compared to the process time constant. It will be discussed later how dynamic differences can be absorbed by controller tuning and implementation choices.

Criteria for Model Selection. The following factors are identified as controller model selection criteria:

- Accuracy of gain prediction
- Accuracy of decoupling and feedforward compensation
- Computational efficiency
- Computational reliability

Nonlinear Algebraic Models. The first model used in GMC to control a distillation column was an algebraic design type model (Cott and Sullivan, 1987). In fact, Cott et.al.(1987) compared several algebraic models for distillation control and found that the Smith-Brinkley (Smith and Brinkley, 1960) model gave the best overall performance. Cott et. al. also examined several other algebraic distillation models including the Jafarey-Douglas-McAvoy model (Jafarey et. al. 1979) and the Smoker equation (Smoker, 1939).

Algebraic distillation models have the advantage that they are, in general, simple to use. For example, the modified Smoker equation (Riggs, 1988b) and the Jafarey-Douglas-McAvoy model can be solved analytically for control action or for parameterization. For systems for which the development of a detailed model is not practical, an algebraic distillation model can be an effective alternative.

Each of the algebraic models assume that the feed tray has the same composition as the feed. This is rarely the case, and as a result, these models do not usually accurately predict the effect of feed flow rate and feed composition changes. In addition, these models assume equimolar overflow and a constant relative volatility each of which may or may not be appropriate to a particular case. In addition, all the algebraic models are based upon a binary system except the Smith-Brinkley model.

We have also found that the application of algebraic models using the relative volatility of the system predicts process gains that are significantly different than the gains exhibited by the process. This can be corrected, at least for a specific operating case, by adjusting the relative volatility used by the model to improve the gain prediction.

For the first application of nonlinear PMBC for distillation control, we used a modified Smoker equation model. We found that the model parameter (the number of theoretical stages) would change over a relative large range of ±25% from day to day. (This relatively large model parameter variation is a direct indication that the model is not very accurate.) The model parameter variations caused by and/or coupled with feed composition changes caused reliability problems for this controller. It required regular controller re-tuning and periodically had to be taken off-control by the operators. The control performance was much improved but the attention required and the reliability were less than desirable.

By an analysis of the assumptions that our algebraic model was based upon and the record of process model/mismatch, we determined that we needed a more accurate column model. This led us to consider how we could use a tray-to-tray binary model as our controller model for distillation column control.

Binary Tray-to-Tray Controller Model. A binary tray-to-tray model was selected because it was expected to overcome the limitations of the algebraic distillation models. The gain prediction accuracy was expected to be improved because of the tray-to-tray modelling and because the feed tray composition is calculated explicitly. It should be noted that the binary model can be adjusted to consider variations in the relative volatility and to consider non-equimolar overflow. In fact, Pandit et.al. (1992) presented laboratory results for nonlinear PMBC applied to a methanol/water column that has both non-equimolar overflow and a factor of six variation in the relative volatility.

In applying the GMC control law, x_{ss} and y_{ss} are determined and the model is used to calculate the boilup rate and the distillate rate, for example. From a degree-of-freedom analysis on a binary column (Luyben, 1989), if the feed composition, feed flow rate, overhead and bottoms products are specified, there is one degree-of-freedom left. As a result, if the boilup rate is calculated from the model, the distillate rate can be calculated by the material balances and then the reflux rate is determined using the equimolar overflow assumption.

81

The solution procedure for the tray-to-tray binary model is as follows. For a specified value of the boilup rate, distillate composition (y_{ss}) and bottoms composition (x_{ss}), the bottoms and distillate product flow rates are calculated by simultaneous solution of the overall and component material balance. Using the equimolar overflow assumption, the reflux rate is equal to the boilup rate minus the distillate rate for a saturated liquid feed. Now the component material balances are solved starting with the reboiler and working up the column to the stage just below the feed stage. That is, knowing the liquid and vapor rates and the composition on a stage and below allows for the direct determination of the composition of the liquid on the stage above. Therefore, starting on the reboiler, one can calculate the liquid compositions up the column.

In a similar manner, the material balances can be used to calculate stage compositions from the top of the column down. That is, knowing the liquid and vapor rates, and the composition of a stage and the stages above allows for the direct determination of the composition of the vapor leaving the stage below. Then using the relative volatility, the liquid composition on the stage can be determined.

For the proper solution, the composition of an internal stage should be the same whether calculated from the bottom upward or from the top downward. Choosing a matching stage near the feed tray results in a well behaved "black-box" function that can be solved iteratively for the boilup rate using the secant method (Riggs, 1988a).

The model parameter for the binary model is the stage efficiency. For the stripping section, the stage efficiency is defined as

$$y_i = x_i + \eta \, (y_i^* - x_i)$$

For the rectifying section

$$x_i = y_i + \eta \, (x_i^* - y_i)$$

where

y_i - is the calculated vapor composition of stage i

y_i^* - is the equilibrium vapor composition of stage i

x_i - is the liquid composition of stage i

η - is the stage efficiency

x_i^* - is the equilibrium liquid composition for stage i

One value of η is used for all stages. Stage efficiencies have been observed to vary $\pm3\%$ over a few days period.

The binary tray-to-tray has proved to be both computationally efficient and highly reliable. Convergence for a 100 tray column on a Perkin-Elmer control computer takes consistently less than one second CPU. In addition, we have demonstrated over 7 column-years experience with this binary tray-to-tray controller model with an outstanding reliability record.

Multi-Component Tray-to-Tray Controller Model. We recognized that the binary model would not be useful for certain multi-component columns. This was due to the multi-component effects upon the composition profiles through the column and to the multi-component effect upon the vapor/liquid traffic through the column. First we tried a multi-component model with fixed relative volatilities. The advantage of this approach is that one does not have to perform a bubble point calculation. Unfortunately, the gain tests against a computer aided design package indicated a poor gain prediction capability. In addition, the numerical solution procedure was not always reliable.

Next we went to the more rigorous multi-component steady-state theta method presented by Holland (1981). In this procedure, there is an inner loop that uses approximate thermodynamic correlations for a simultaneous solution of the material and energy balances for all the stages in the column. The outer loop involves updating the empirical thermodynamic correlations using the rigorous vapor/liquid calculations.

The inner loop is based upon the theta method (Holland, 1981). In the theta method, by neglecting the effect of composition on component K-values, the molar flow rates of each component are determined by the solution of a tri-diagonal set of linear equations. The value of theta is calculated based upon satisfying the overall component material balances. Then a correction factor, which is determined from the value of theta, is applied to all the vapor and liquid compositions. It was found that the theta correction tended to over-correct the stage composition estimates particularly for the stages around the feed stage. A small amount of filtering on the corrected stage compositions improved convergence speed and convergence reliability.

Since the GMC control law provides x_{ss} and y_{ss}, the steady-state multi-component column model must be solved several times (usually 4 to 8 times) in order to determine the boilup rate that satisfies x_{ss} and y_{ss}.

The model parameter is the Murphree stage efficiency (Van Winkle, 1967). Some modification of the equations used to implement the theta method is necessary to include the Murphree stage efficiency. One value of the Murphree efficiency is applied to all stages and all components.

The multi-component controller model has proven so far to be computationally reliable and reasonably computationally efficient. We have checked convergence of the multi-component model over a wide range of operating conditions and as a part of the controller tested on a dynamic simulator of a depropanizer and in both cases have found it to be highly reliable. Controller implementation requires four to eight steady-state solutions which requires about 5 to 10 seconds CPU on a 33MHz 486 PC.

4.4 IMPLEMENTATION ISSUES

Dynamic Column Modeling. For each of the nonlinear PMBC distillation control projects, we have developed a dynamic model of the column and used the dynamic model to test the nonlinear PMBC controller. This greatly reduces the amount of on-process development time. That is, a careful job of dynamic modeling results in a dynamic model that has much of the true dynamic character of the real process and as a result presents many of the control challenges the real process offers. In addition, if unexpected

85

problems develop (e.g., identification of an unexpected operational constraint), the dynamic column model allows one to quickly analyze the problem, test possible solutions, and arrive at a workable solution. Finally, the tuning of the controller used on the dynamic simulator provides good estimates of the tuning parameters for the actual column.

A major decision in the development of a dynamic column simulator is what type of vapor/liquid equilibrium (VLE) model to use. For the binary model, the VLE is conveniently represented using a relative volatility. For the binary cases that we considered, a constant relative volatility was a reasonable assumption. For the multicomponent model, we used the Souave-Redlick-Kwong (SRK) (Holland,1981) model. The SRK model is computationally intensive requiring two iterative searches for each K-value evaluation. As a result, we found it to be computationally advantageous to use the K_B model (Holland, 1981) for bubble point calculation and to calculate the K_B model parameters using the SRK algorithm. Specifically, K_B models for all stages were reparameterized each 30 seconds of simulation and the correlation for a tray was parameterized before 30 seconds of simulation time if the temperature of that tray changed by more than 1°C since it was last parameterized. This provides a major speedup of the dynamic model resulting in a dynamic model that was 60 times faster than real time for a 39 stage, seven component column model run on a 33Hz 486PC.

The following items pertain to both binary and multi-component dynamic column modeling.

- **Effective molar holdup of each stage.** A tray in a distillation column is a distributed parameter problem yet it is usually modelled as a perfectly well mixed vessel. As a result of the transport time for the liquid through the downcomer and the transport across the tray, the effective molar holdup will be at least as large as the holdup of liquid on the tray and the liquid holdup in the downcomer. The more the flow through the downcomer and the flow across the tray approaches plug flow, the larger the effective molar holdup will be. As an upper limit, one can expect an effective molar holdup as much as twice the actual sum of the downcomer and tray holdup if perfect plug flow existed. Since the effective molar holdup has a significant effect upon the predicted composition dynamics of a dynamic column simulator, when one underestimates the effective molar holdup of the tray, the simulator will predict faster composition dynamics than would be expected on the real process. Also the molar holdup in the accumulator and the reboiler should be accurately estimated.

- **Level control dynamics.** The dynamics of level control can have a very significant effect upon the dynamics of a column. For example, very loosely tuned level controls can result in composition dynamics that are inverse acting. As a result, it is important to accurately model the level controls of a column. The level indication of 0 to 100 percent on an accumulator may mean that it is empty when 0 and full at 100 percent or it could mean that it is 40% full at 0 percent level and 60% full at 100

87

percent.   Therefore care should be taken to model the level control process accurately.

- **Analyzer delays.**   The total product composition and feed composition delays can have a dominant effect upon the dynamic behavior of a column particularly under feedback control.   The total analyzer delay is the transport delay from where the sample is taken until it arrives at the analyzer plus the analyzer update frequency.

- **Valve dynamics and heat transfer dynamics.**   For relatively fast acting columns, valve dynamics and heat transfer dynamics should be modelled.   Even though the time constant is only of the order of three to five seconds, it can have a significant effect upon controller tuning results.

- **Inferential estimates of product composition.**   When tray temperatures are going to be used to estimate product composition, the inferential approach should be implemented on the simulator as well.

- **Constraint modeling.**   The various constraints on two product distillation columns result in a limitation on the boilup rate.   The trick is to use process data to correlate the constraint to a limitation in boilup rate.   In this manner, constraints can be directly added to the dynamic column simulator.

Constraint Handling.   As indicated earlier, three of the four high reflux ratio columns experience condenser duty constraints during the hot summer days. During these periods, the energy input to the column is set in order to

88

maintain the reflux temperature at a maximum level. Since the energy input is set independently, the only degree-of-freedom left to manipulate is the relative production rates of the products. Because the distillate product is a saleable product with hard constraints on the product purity, the distillate rate was adjusted to maintain the overhead product purity while the bottom product purity was allowed to float.

The identification of the constraint is based upon reflux temperature while a comparison of the constraint control selected energy input with the energy input calculated from the dual composition control nonlinear PMBC control is used to determine when one can return to unconstrained control. If care is not taken when a column is operating near the constraint, it can cycle into and out of constraint control over a period of time. This problem can be overcome by making it harder to leave constraint control than to enter it.

Bumpless Transfer and Rate of Change Limits. Bumpless transfer allows for a smooth transition into the nonlinear PMBC controller while rate of change limits prevent the controller from making excessively large changes in the manipulated variables. There are at least a couple of ways to approach bumpless transfer. For the industrial column considered here, the objective of bumpless transfer is to be able to turn on the nonlinear PMBC controller during a major upset and have the controller stabilize the process and drive it to its respective setpoints. Fortunately, the controller models are quite good the old model parameters are usually sufficiently accurate to allow the controller to take over "cold". In addition, placing rate of change limits on the manipulated variables (e.g., steam flow rate and distillate rate) is a major

89

factor in the stable operation of this approach to bumpless transfer.

When the model parameters are likely to have changed significantly since the last time it was used, a different approach can be used. For this case, it is important that the column is relatively stabilized but not necessarily at the desired setpoints. Then when the controller is turned on, the controller model is parameterized to match the current operation and the setpoints used by the controller are set equal to the current measured values of the product compositions. In this manner, the manipulated variable values remain unchanged for the first control interval. Then the controller setpoints are ramped to the actual setpoints at a preselected moderate rate.

Rate-of-change-limits restrict the movement in the manipulated variables. This is certainly desirable during analyzer failure, erratic behavior of the analyzer, or if the operator fails to turn off the controller while the analyzer is being calibrated. In addition, the effects of certain operational problems or conditions can be negated by the use of rate of change limits.

Overall, well conceived bumpless transfer and rate of change limits can enhance the reliability of a nonlinear PMBC controller and likewise enhance its acceptance by operations.

Operator Acceptance. A major factor in any successful advanced process control project is operator acceptance. No matter how technically successful a control project may be, it will not be an economic success unless the plant operators are willing to use it. Even if improved control is demonstrated which subsequently makes their job easier, they may not want to use the advanced control because of one or more of the following concerns:

-       They are afraid that advanced computer-based control will cost them their jobs. After all, a large part of their job is "riding" the column through tough times and if new controls eliminate or greatly reduce that need, the company might not need them anymore.

-       Improved control will just lead to tighter product specification which may even result in more difficult work for them.

-       The controller may make changes that are counter to their experience. For example, an operator will typically increase the reflux rate to reduce the impurity in the overhead product (one-dimensional approach) while the nonlinear PMBC controller will usually make changes in both the boilup rate and the reflux rate even if the bottoms product is on specification (a two dimensional approach). This difference will worry some operators and cause them to question the reliability of the controller. In addition, the nonlinear PMBC controller will typically make sharper and larger changes in the manipulated variable than the operators are used to.

-       They do not usually like new ways of doing things. It's human nature to resist change that is imposed upon you.

In order to overcome these and other problems associated with operator acceptance, communication lines between the operators and the process control engineers on the project must be opened early and maintained even after the advanced controller has been commissioned and turned over to operations.

Moreover, it is essential to get as much input from the operators as possible before the controller is implemented. The operators are a valuable source of operating history and can provide information about operational objectives, process constraints, and abnormal operations that is essential to the formulation of a reliable and effective controller. If the controller is not designed to handle their real problems, it will be of little use to operations.

It will be much easier to achieve operator acceptance if they feel that they are part of the project and that it is not being imposed upon them. Asking for their input before the project begins will help in this area. Also, explaining the economic incentive of improved control can help. Explaining in their terms how the controller works will help alleviate some of their fears. For example, they should be made aware that the controller is likely to make changes that are sharper and different from what they are used to. They will not, in general, trust a "black box" controller.

In summary, developing a personal relationship with each operator based upon communication should enhance your chances of having them feel that they are a significant part of an advanced control project. Then when improved control can be demonstrated, it should make their job easier while allowing them to operate their unit more economically efficient. In a truly successful control project, the operators, the control engineers, and management should each feel that they have made a contribution to the project and that the project has been a success from their point of view. Along this line, it is essential that management, operations, and the control personnel agree at the beginning to the operational objectives of the project.

4.5 RESULTS

Tuning Procedures. Tuning a nonlinear PMBC controller is similar to field tuning a PI controller but since the proportional gains are dimensionless, initial estimates of tuning parameter values are easier to make since we had experience tuning the controller for the dynamic simulators. By running simulations, we have found that the greater the analyzer dead-time, the less proportional gain one should use. Also, in general, the larger the process time constant, the larger the proportional gain that can be used.

For each industrial case, the nonlinear PMBC controller was tested on a dynamic simulator of the column. As a result, we have a good idea of the tuning parameters before application of nonlinear PMBC to the actual column has been started. From experience we have found that proportional tuning parameters (K_{11} and K_{12}) that work best on the actual column are about 20% lower than the optimal tuning parameters developed on the dynamic simulator.

For the low relative volatility columns, proportional gain tuning parameters ranged from 2.0 to 6.0 with the larger time constant process requiring the larger proportional gains. In each case, the tuning parameters are fine tuned by observing the measured variables and the manipulated variables action.

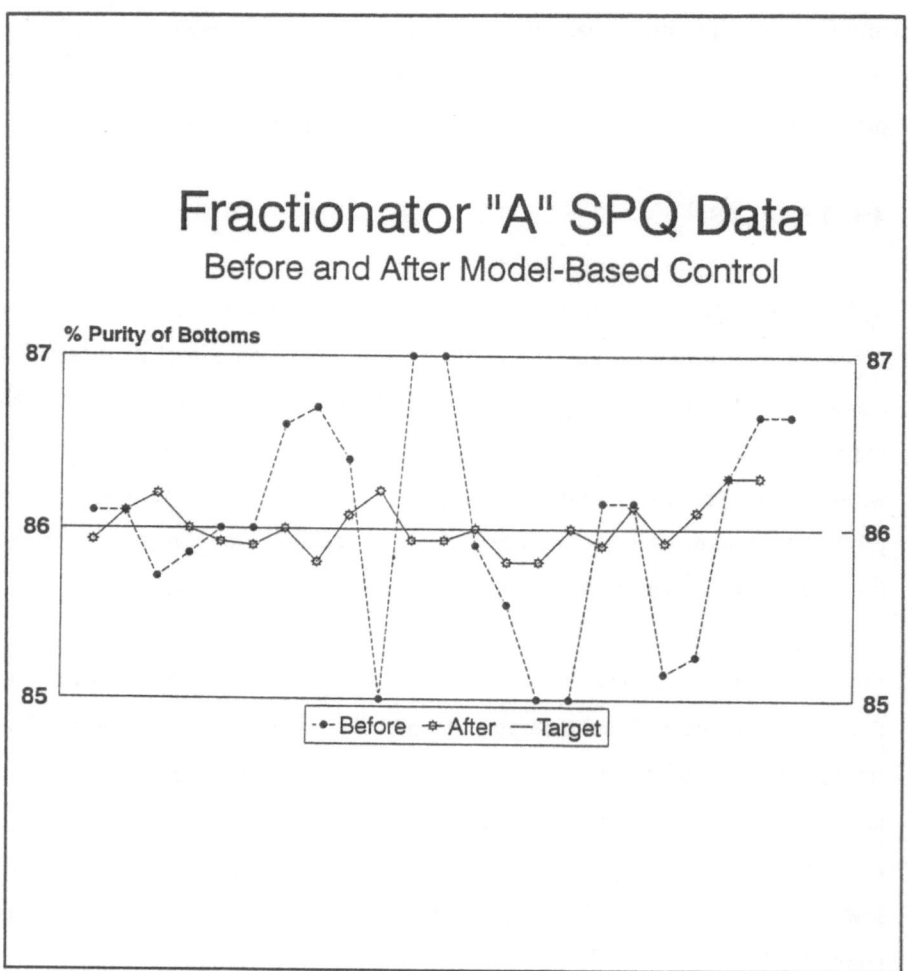

Figure 4.3   Fractionator "A" SPC Data

That is, if the process is slow responding to an upset, then the proportional gain is increased.   Or if ringing in the controlled variable or manipulated variable is observed, then the proportional gain is decreased.

<u>**Control Performance for Low Relative Volatility Columns.**</u> At this point, nonlinear PMBC has been applied to four high reflux ratio binary-type columns with a total of over 7 column-years of total operating experience. Figures 4.3 and 4.4 show week duration SPQ charts of product uniformity for

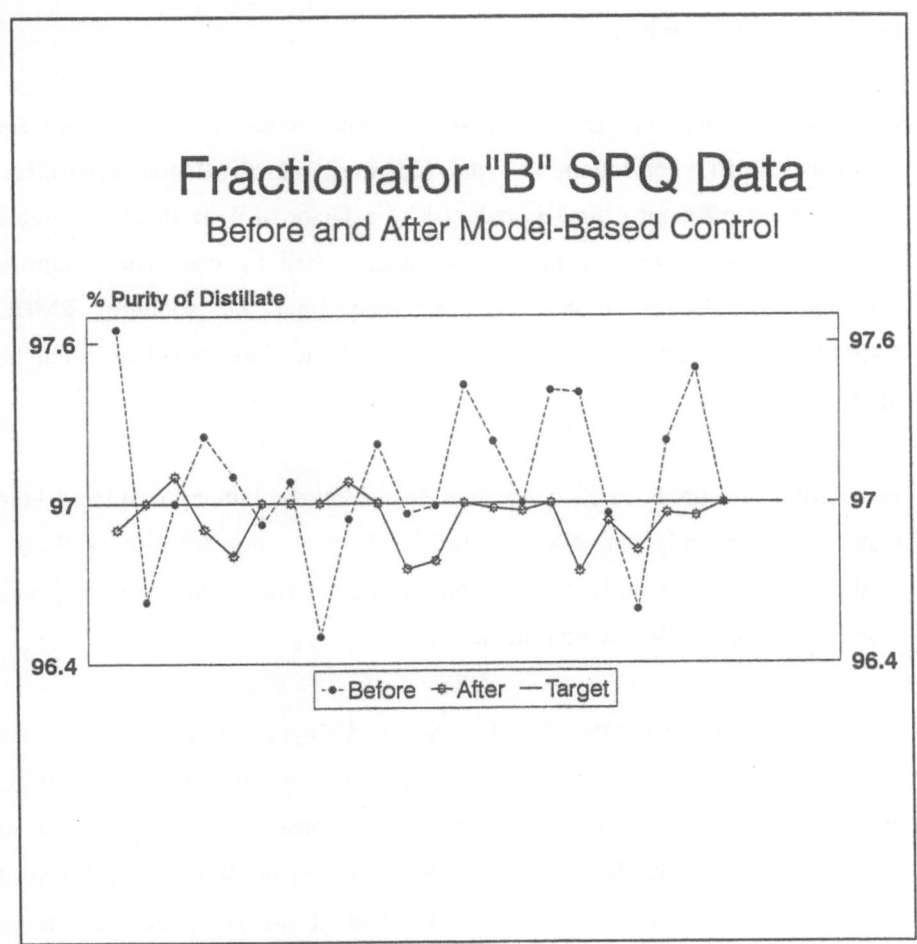

Figure 4.4 Fractionator "B" SPC Data

comparison between the nonlinear PMBC controllers and the previous single loop PI controllers. Each column is responsible for producing one final product. In each case, the results were chosen to be representative of normal operation of both controllers. Note that the variability in the products produced by these columns is reduced by a factor of 2 to 5 compared with the PI controller results.

Figure 4.5 shows a comparison of on-control time versus off- control time for nonlinear PMBC controllers and the remaining PI distillation controllers. Note that the off-control time is reduced by a factor of 4. It should be noted that the off-control time includes downtimes caused by mechanical failures and analyzer failures. It has been estimated that the nonlinear PMBC controllers are being used greater than 95% of the time that they could be used.

In addition, the previous PI controllers on the four columns considered here required almost daily tuning and a high level of operator attention while the nonlinear PMBC controllers are rarely re-tuned after commissioning and require relatively little operator attention.

Control Performance for the Multi-Component Column. Figure 4.6 shows the impurity levels for the top and bottom products for the nonlinear PMBC applied to the dynamic simulator of the depropanizer for a ramp change in feed composition. For this case, the mole fraction of the light key in the feed (propane) was decreased by 5 mole % while at the same time the mole fraction of the heavy key in the feed (iso-butane) was increased by 5 mole %

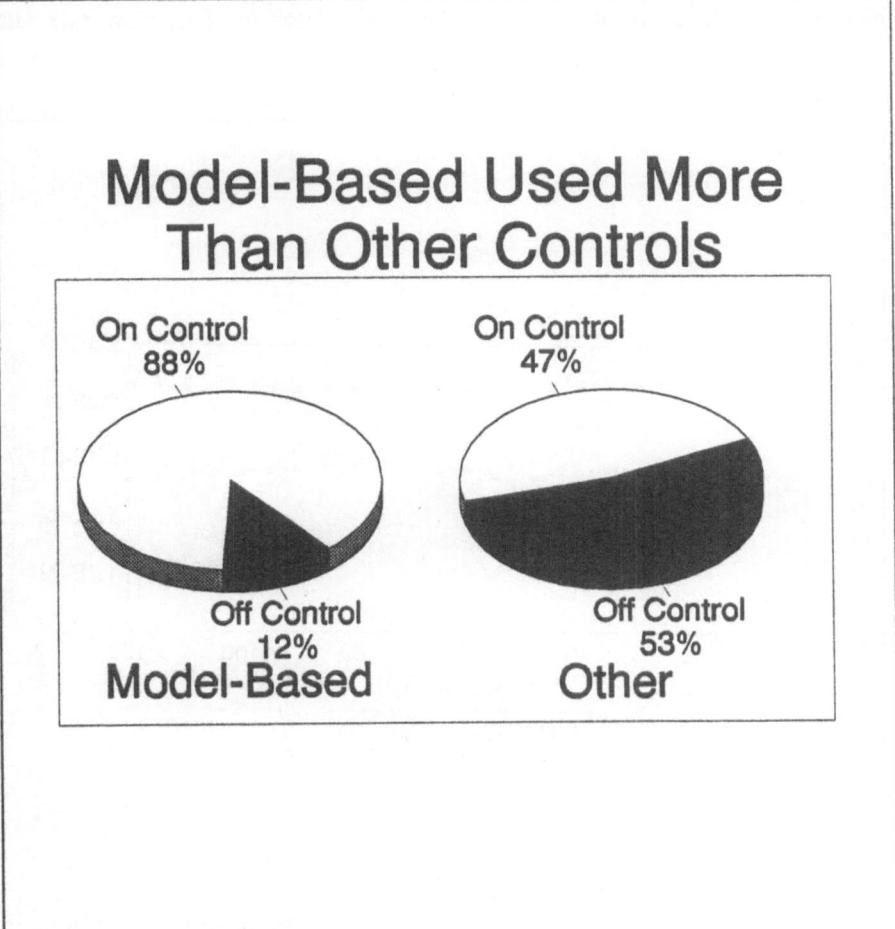

Figure 4.5   Percentage time on-control

over a 30 minute period.

Figure 4.7 shows the impurity levels for the top and bottom products for a 20% step increase in feed flow rate to the column.   The nonlinear PMBC controller produced results on the column simulator for feed flow rate and

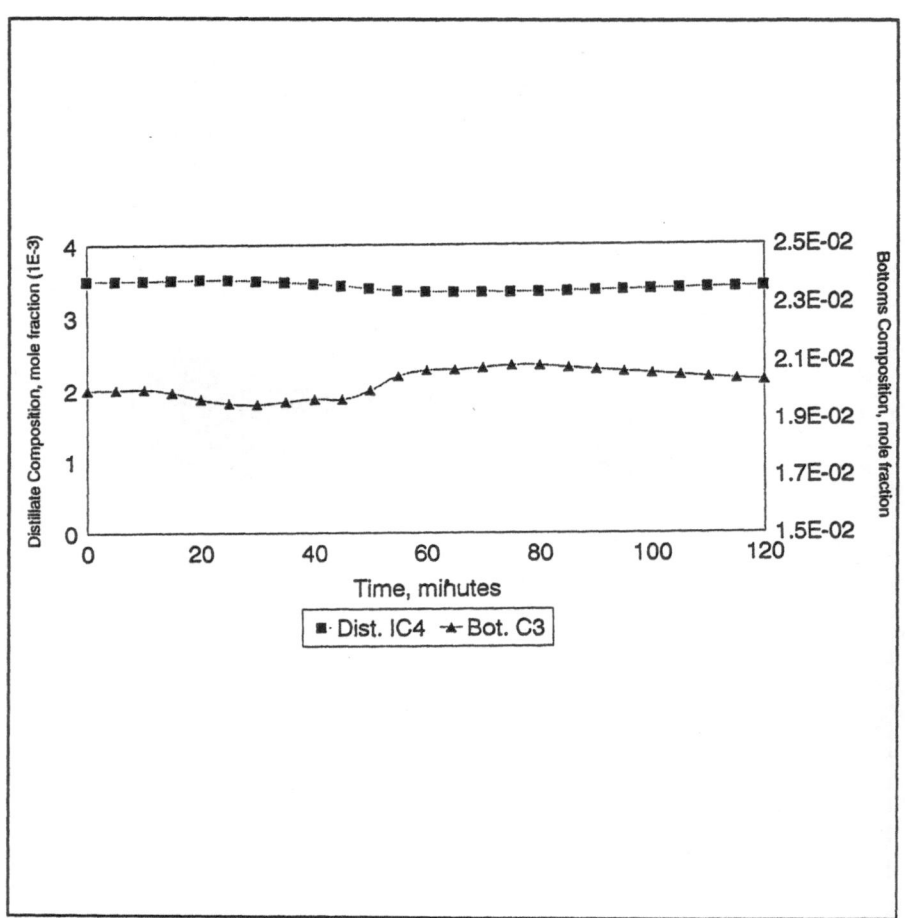

Figure 4.6   Model-Based Control with Feed Composition Change

feed composition upsets that are far superior to the control performance of
the existing PI controls on the actual depropanizer.

4.6 ECONOMIC BENEFITS

The improved control of each of the four low relative volatility columns has resulted in significant economic benefit and very short project pay-back periods.

For the first column that nonlinear PMBC was applied to, approximately 30% of the product previously produced by this column had been off-specification thus requiring re-processing. After installation of the nonlinear PMBC controller, the production of off-specification product was virtually eliminated. As a result, approximately 30% less feed had to be fed to this column and the proceding column in order to meet production quotas. Therefore, the reboiler steam requirements were reduced by 30% resulting in an energy savings of about $US250,000 per year. In addition, an increase in production capacity of 30% is also available if needed. The second and third columns that had nonlinear PMBC applied to them were in similar service. For both cases the improved control made it possible to reduce the amount of over-refluxing while producing a more uniform product. The energy savings amounted to a 20% reduction in steam usage yielding an annual savings of about $US450,000. In addition, greater processing capacity is also available as a result of improved control.

The fourth column has been identified as a bottleneck in the production of a product for which the refiner can sell all that it produces. The improved control on this column has resulted in a 15% increase in product rate which corresponds to a $US2,000,000 increase in profits annually.

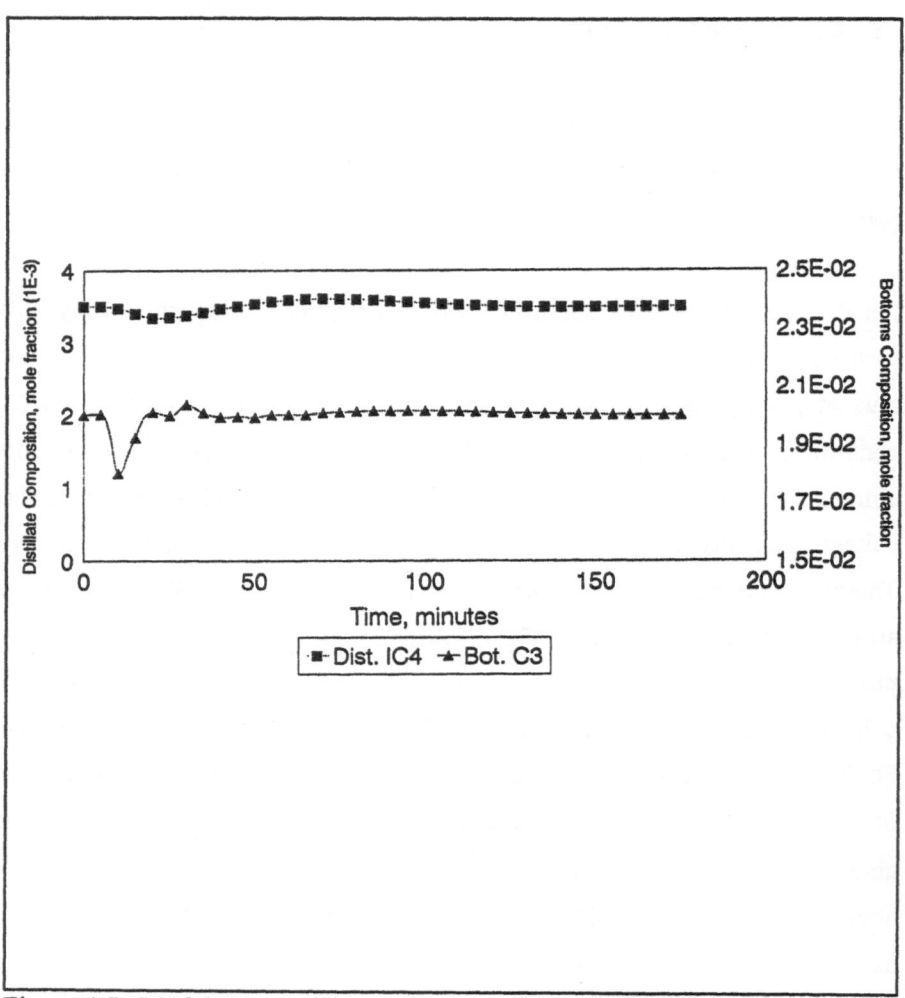

Figure 4.7 Model-Based Control with Feed Rate Change

4.7 CONCLUSIONS

The use of nonlinear models for distillation columns compensates directly for the inherent nonlinearity of columns and provides accurate decoupling and feedforward compensation. An accurate controller model can predict the nonstationary behavior of the process gains as the result of changes in operating conditions.

As a result, nonlinear PMBC has been successfully applied to the control of four low relative volatility industrial columns. Factors of 2 to 5 reduction in the product variability have been demonstrated while the nonlinear PMBC controllers have a service factor in excess of 95% and require much less operator and control engineer maintenance and tuning time. Since the projects reported here require only software changes, the pay out periods have been one month or less in each case.

4.8 REFERENCES

Bhat N. and McAvoy T.J. (1990) Use of Neural Nets for Dynamic Modeling and Control of Chemical Process Systems.   Computer Chem Engr 4:517-532

Cott B.J. and Sullivan G.R.   (1987)   Process Model Based Control and Optimization of a Binary Distillation Column, presented at the Spring National AIChE Meeting, Houston, TX

Cott B.J., Reilly P.M. and Sullivan G.R. (1987) Selection Techniques for Process Model Based Controller, presented at the Spring National AIChE Meeting, Houston, TX

Cutler C.R. and Ramaker B.L. (1979) Dynamic Matrix Control-A Computer Control Algorithm. AIChE 86th National Meeting, also in Joint Automatic Control Conf. Proceed, San Francisco, CA

Economoric C.G., Morari M. and Piasson B.O. (1985) Internal Model Control 5 Extension to Nonlinear Systems. Ind Eng Chem Process Des Dev 25:408-418

Holland C.D. (1981) Fundamentals of Multicomponent Distillation, McGraw-Hill

Humphrey J.L., Seibert A.F. and Koort R.R. (1991) Separation Technologies-Advances and Priorities, US DOE Report, DOE/10/12920-1.

Jafarey A., Douglas J.M. and McAvoy T.J. (1979) Short Cut Techniques for Distillation Column Design and Control. Ind Eng Chem Process Des Dev 18:197

Lee P.L. and Sullivan G.R. (1988) Generic Model Control. Comp. and Chem. Eng. 12:573-583

Luyben W.L. (1989) Process Modeling Simulation and Control for Chemical Engineers. 2nd edition, McGraw-Hill

Pandit H.G. (1991) Experimental Demonstration of Nonlinear Model Based Control Techniques on a Lab-Scale Distillation Column, PhD Dissertation, Texas Tech University, Lubbock, TX.

Pandit H.G., Rhinehart R.R. and Riggs J.B. (1992) Demonstration of Constrained Process Model Based Control of a Nonideal Distillation Column, Proceed of the 1992 American Control Conf. Chicago, IL.

Parrish J.R. and Brosilow C.B. (1988) Nonlinear Inferential Control AIChE J 34:633

Riggs J.B. (1988a)   An Introduction to Numerical Methods for Chemical Engineers. Texas Tech University Press

Riggs J.B. (1988b) Nonlinear Process Model Based Control of a Propylene Sidestream Draw Column. I&ECR 29.

Smith B.D. and Brinkley W.K. (1960)   General Short Cut Equation for Equilibrium Stage Processes. AIChEJ 6:446

Smoker E.J. (1939)   Analytic Determination of Plates in Fractionating Columns. Trans AIChE 34:165

Van Winkle M. (1967) Distillation. McGraw-Hill, New York USA

CHAPTER 5

AUTOMATIC MOISTURE CONTROL IN PARTICULATE DRYERS

5.1 ABSTRACT

Although drying is one of the oldest and most common unit operations in the process industries little work has been done on the automatic control of process dryers. This is largely because of the complex nature of drying processes and lack of appropriate process control software and hardware including moisture sensors.

This chapter presents the state-of-the-art of process control of particulate dryers including the incentives for process control, the required sensors and computer control hardware and software. A case study of a continuous horizontal conveyor belt dryer for drying pet food is used as an illustration.

The work included the development of a moisture sensing system capable of measuring the full range of products produced in a typical pet food plant, the GMC controller and implementation on an industrial dryer. The control system manipulated the temperature of the final two drying zones to control the drying rate, in order to maintain the moisture content at the desired level.

The system demonstrated the ability to reduce overdrying associated with manual control, resulting in a decrease in the energy consumption and an increase in production yield. The actual economic benefits depend on dryer characteristics and product throughput.

Conclusions of this case study and many others like it show that, although there are significant economic benefits from the computer control of dryers, the non-linearities in the process as well as the difficulty in process measurements make this a challenging class of processes to control.

5.2 INTRODUCTION

Moisture control during drying is important because it relates directly to the economic viability of the drying operation. If the product to be dried fails to meet moisture specifications several consequences may result depending on the product and the reasons for drying. For example in the case of grains and food stuffs underdrying may result in spoilage and overdrying results in increased energy costs and reduced yields. In the case of plastics underdrying may result in a loss of product performance and will reduce the processability in subsequent processing steps.

Dry pet food is produced by extruding a mixture of grains, meat and meat by-products with water. Steam is added to the barrel of the extruder to adjust the expansion of the pellets as they leave the extruder. This allows the operators to control the bulk density of the product. This is important in a manufacturing environment when a certain weight of product must fill a specific sized bag.   The wet product is fed evenly onto the conveyor belt by an oscillating spreader.   The belt consists of perforated metal plates which allow air circulation through the product.   The dryer normally consists of at least two heating zones.   In a heating zone, air is circulated through the product layer on the belt.   A fraction of the air is vented to the discharge system.   The remaining air flow is mixed with fresh air and passed by a heater (typically steam coils or a natural gas burner) in order to maintain the desired air temperature.   A centrifugal fan is used to circulate the air through the layer of product.   The last zone in the dryer may be used for product cooling or as a final drying zone.   Variations of this design include multiple conveyors in the dryer which have the effect of increasing the dryer residence time without increasing the dryer size.

The extruded pellets have an initial moisture in the range of 20 to 25% w.b. The moisture level must be lowered below 12% w.b. to ensure that spoilage will not occur in the finished bagged product.   Typical product flows through a dryer are 5 tonnes/hour.   This results in an evaporation load of about 0.67 tonnes/hour.   These dryers typically operate at about 50% efficiency.   This means each dryer will require 12.2 GJ/h to meet the energy requirements. The flowrate of natural gas consumption to provide this energy would be about 360 m^3/h.

A schematic diagram of a conveyor belt pet food dryer is shown in Figure 5.1.   The moisture content of the outlet product is often difficult to control because of the many changes that occur to the inlet product and the drying conditions with time.   For example, the moisture content of the inlet may vary considerably with time as the operators adjust extruder conditions to maintain product size, shape and bulk density. The drying rate can also change with time as well as the ambient conditions such as temperature and relative humidity. These changes make it difficult to consistently maintain the outlet moisture content to a specific value.

Previous methods to control moisture relied on operator experience to guide the operation of dryers because on-line sensors were not available to continuously measure moisture content. On-line moisture sensors are now available, (Carr-Brion (1986), NIR Systems (1989)), and have given the dryer operator the potential to control the outlet moisture much closer to target. The performance of these systems, however, needs to be examined closely. For example, it is possible for an operator to actually reduce the performance of a dryer by the inappropriate adjustment of throughput or temperature in response to changes in inlet moisture compared to the case where the operator makes no changes at all. This is also possible with computer control if care is not taken in the tuning of the controller. Also, because these sensors are not always accurate, frequent calibration may be required to maintain reliable comparison with manual reference meters (Reilly et al, 1988).

Dryers in the pet food industry are normally controlled based on results of hourly manual samples. Drying air temperatures are then adjusted based on these results.

The operators control the dryer by taking hourly samples and analysing for

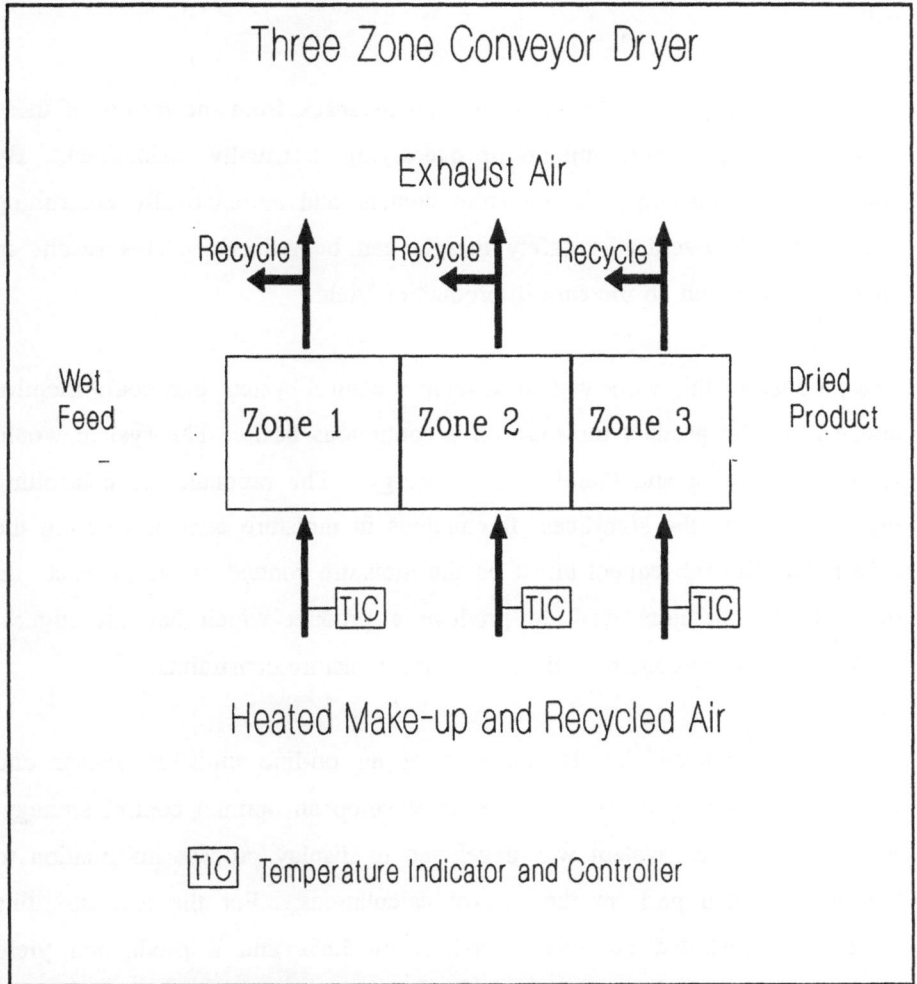

Figure 5.1 Schematic Diagram of a Conveyer Belt Dryer

moisture content in the laboratory. The drying air temperatures may then be adjusted as required. Operators tend to overdry due to a lack of on-line moisture feedback from the dryer and typically average between 7 and 8%

w.b.

Since operators generally have no on-line feedback from the results of these changes, a large safety margin of overdrying is usually maintained. By continuously monitoring the moisture content and automatically controlling the dryer, this overdrying safety margin can be reduced. This results in energy savings and an increase in production yield.

The purpose of this work was to develop a control system that could monitor and control the product moisture on a continuous basis. The system would reduce overdrying and therefore save energy. The rationale for controlling dryers is due to the significant fluctuations in moisture content entering the dryers, and the subsequent affect on the moisture content of the product. In many cases operators wish to produce a product which has the highest possible moisture content without violating moisture constraints.

This work involved the development of an on-line moisture sensor and monitoring dryer operations in order to develop an optimal control strategy. A computer based system was developed to display process information to the operator and perform the control calculations. For the test site, this system demonstrated an energy savings of 5.6% and a production yield improvement of 1%. This resulted in an economic benefit of $CDN123,600 per year based on the energy savings, yield improvement and continuous monitoring benefits.

5.3 RATIONALE

Pet food dryers are typically poorly instrumented for continuous moisture control. Operators have access only to drying zone temperature readings as shown in Figure 5.1. In order to obtain information about the product, operators must physically obtain a sample and analyse it for moisture content. Due to this lack of information, operators tend to overdry the product. Product that is too wet will be rejected because the chance of spoilage is high. Bulk density, an important product property, is controlled by adjusting water and steam injection flows in the extruder barrel. This results in changes in the inlet moisture of the feed to the dryer which create disturbances in the drying process and another level of uncertainty for the operator.

An automated control system will provide continuous on-line monitoring of product moisture from the dryer and will be able to quickly adjust the dryer operation to compensate for changes in outlet moisture content. The control system would also display all the measured process parameters. This provides the operator with information so that the process can be operated more effectively. Since product moisture levels will be controlled, the average moisture content can be controlled closer to the maximum allowable limit. The higher moisture obtained will result in improved product yield, better product uniformity, and save energy.

Pet food drying was an attractive market to produce a control system for since there are no known computer control systems commercially available and the pet food industry in North America has a market value in excess of seven billion dollars U.S. per year. Several on-line moisture sensors are available but they tend to be too expensive or do not provide the user with enough calibration curves required for the large number of products in a pet food plant.

5.4 RESEARCH AND DEVELOPMENT OF THE CONTROL SYSTEM

5.4.1 Moisture Control

Several dryer manufacturers supply control systems based on product temperature or exhaust air temperature (Behlen (1986), LAW (1986)). The operator must select an exhaust air temperature or grain bed temperature resulting in a desired outlet moisture content. This may be difficult and will change with grain variety, inlet moisture content and atmospheric conditions. Success of these controllers is limited since they have no feedforward capabilities and the relationship between temperature and outlet moisture is not constant. The relationship between outlet moisture and outlet temperature is a function of the inlet moisture, inlet air temperature, humidity and physical properties of the product. In general, the higher the air temperature is, the dryer the product will be. However, because this relationship is complex it is difficult to use in a feedback or feedforward controller.

Control systems which monitor both the inlet and outlet moisture levels should provide superior control of the drying process. A control system developed by Shivvers, was tested by Michigan State University (Bakker-

Arkema et al, 1988). This system measured the moisture content of the inlet and outlet grain using a discrete sampling system with a sampling period of 3-10 minutes. The control algorithm was a feedforward type with feedback correction and dynamic compensation in which the throughput rate was manipulated to control the outlet moisture (Moreira and Bakker-Arkema, 1990).

Model-based control has become an area of much interest in the last decade. Lee and Sullivan (1988) review the current state of the art in this area as well as present a new framework for model-based control call Generic Model Control (GMC). The basic concept behind model-based control is that a model of the process is used as a guide to controlling the actual process. With a knowledge of the inlet moisture, outlet moisture target and a model of the dryer the throughput rate can be adjusted accordingly to control the outlet moisture.

Figure 5.2 shows the outlet moisture distributions from a pet food conveyor belt dryer using two different control strategies. In all cases the outlet moisture was controlled by adjusting the dryer temperature by manipulating the valve position on the natural gas line to the burner. The moisture content at the highest frequency is the mean moisture and the spread or width of the distribution represents the standard deviation or variation in the outlet moisture. The vertical line at 11.5% indicates the maximum desirable moisture content. Good moisture control is represented by a narrow distribution about the mean close to, but lower than 11.5%. Poor control, on the other hand, is represented by a broad outlet moisture distribution either

113

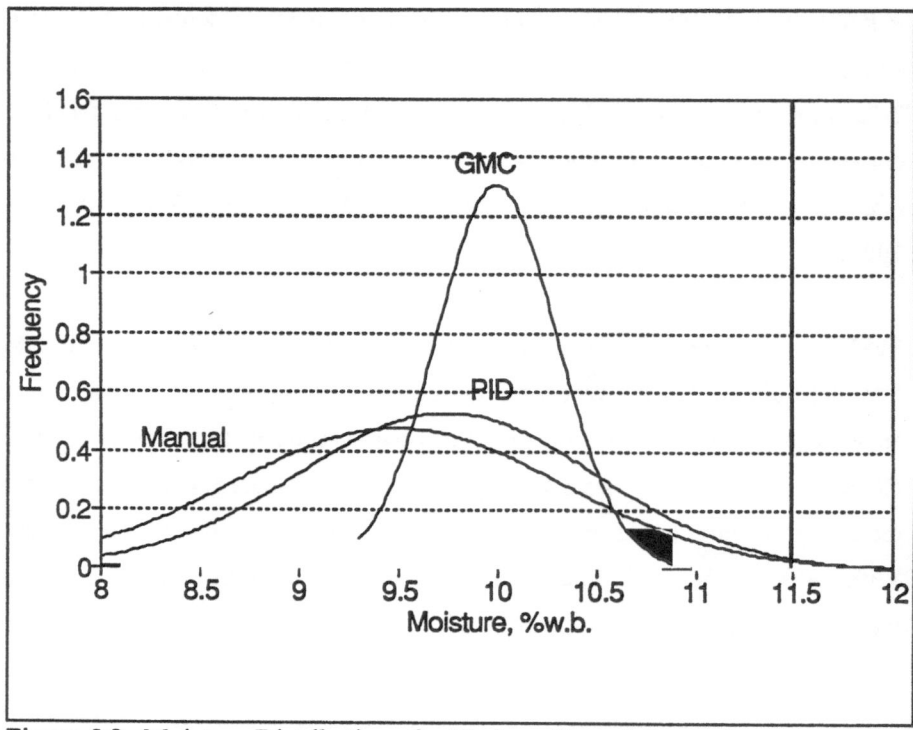

Figure 5.2   Moisture Distributions for Various Control Strategies

above or below the target of 11.5%.

If one assumes that manual control represents the base-line operating conditions then the standard PID (Proportional Integral Derivative) control algorithm results in an improvement over manual control. The GMC model-based controller in turn results in an improvement over PID control. This is expected because an explicit model of the dryer is built into the GMC controller. The model-based control strategy with model update has been incorporated into the Dryer Master (DM) moisture control system, (Dantec,

1992).

With the development of a moisture metering system to measure the trends in product moisture content, it was then possible to develop a control strategy. For the dryer in the study, there were four possible manipulated variables; the conveyor belt speed and the temperature in each of the three drying zones. The effect of changes to each of the four variables was studied by conducting open loop tests during dryer operation. The results were used to configure a simulation of the dryer in order to fully understand the dynamics of the drying process in this case.

5.4.2   Conveyor Belt Speed

The conveyor belt speed determines the residence time in the dryer and therefore could be used to control the moisture. During the open loop tests however, changes in the conveyor belt speed seemed to have unpredictable results. For example, a relatively small increase in speed may result in a decrease in moisture content while a larger increase in speed may show an increase in the moisture content. These observations were simulated using a rigorous simulation program and the results indicated that as the conveyor belt is slowly increased, the bed height drops and the air recirculation rate increases, increasing the drying rate. This increase in drying rate initially increases proportionally faster than the decrease in residence time of the product in the dryer. This results in the observed decrease in moisture. A critical point is reached where the drying rate stops increasing and the residence time determines the moisture content and the moisture therefore begins to increase.   Similar observations were made for decreases in the conveyor belt speeds.

Since it would be difficult to determine during actual operation where the dryer operation was relative to the two critical points, the conveyor belt speed is not suitable for accurate moisture control.

5.4.3   Zone 1 Temperature

In the first temperature zone, the product contains mostly free moisture since it is the dryer section immediately after the extruder.   The drying rate in this zone is determined by the maximum rate of evaporation from the surface and is not influenced by small changes in zone 1 temperature.   Increases in temperature large enough to influence the rate would create a dry crust on the outside of the particle which would affect further drying and processing, such as fat absorption.   Decreases in the temperature of zone 1 large enough to slow the evaporation rate will allow the product to cool and therefore slow the rate of evaporation in the other zones since the water transport mechanisms to the surface would be slowed until the particles are reheated.

Manipulation of the zone 1 temperature is not suitable for moisture control and it will be left to the operators to determine a suitable constant operating temperature, based on previous dryer operation and the current outlet moisture reading.

5.4.4   Zone 2 Temperature

Since the free moisture is largely evaporated in zone 1, the drying rate in zone 2 is determined by the surface evaporation rate and internal moisture transport mechanisms. The air temperature in zone 2 will determine the overall drying rate since it will control both the surface evaporation rate and the temperature of the product. The product temperature will directly control the internal water transport mechanisms.

Although the results of changes to the air temperature of zone 2 can be predicted the accuracy is not sufficient for control since the moisture content entering zone 3 can influence the drying rate in that zone and therefore the final product moisture content.

5.4.5   Zone 3 Temperature

The drying rate for zone 3 is determined in the same way as zone 2. The effect of changes in zone 3 temperature can be predicted more accurately than for zone 2 since there is no further processing between zone 3 and the outlet moisture meter. For example, a 15°C change in zone 3 will result in about a 1% change in product moisture content.

Since the results of changes in zone 3 resulted in the most easily predictable change in moisture content, it was decided to manipulate the air temperature of zone 3 to control the drying rate in order to achieve the desired moisture levels.

5.4.6   Dryer Control Modifications

In order for the computer control system to manipulate the air temperature of zone 3, it must be able to communicate with the single loop controllers used for each temperature zone.   Two options are available for such communications; digital (RS422 or RS232) and analog (current or voltage). The digital communications protocol will vary among manufacturers of single loop controllers.   Therefore, for this system two current loop communications were specified.   The temperature controller for each zone will send a current signal proportional to air temperature to the control system.   The control system will in turn send a current signal proportional to air temperature setpoint as requested by the operator or calculated by the control algorithm back to the temperature controller.   A schematic showing all instrumentation locations and analog signals is shown in Figure 5.3.

5.5 DEVELOPMENT OF THE CONTROLLER

A Dryer Master system has been developed to control the drying of pet food in a conveyor dryer.   The control system consists of a centralised computer system which receives incoming data, presents data to the operator, performs control calculations and sends set points to existing single loop temperature controllers.

A radio-frequency, capacitance type moisture meter monitors the product exiting the dryer.   A fraction of the total product flow is diverted into a containment vessel.   The moisture meter is mounted in this vessel along with

118

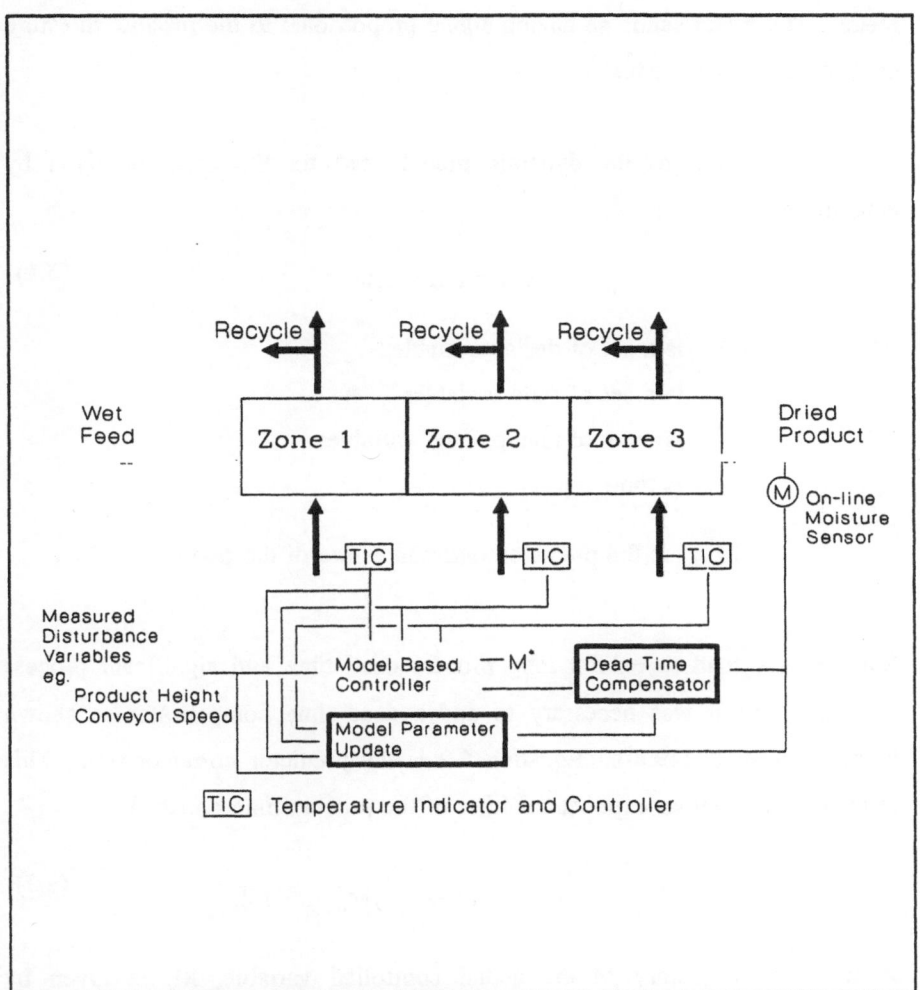

Figure 5.3 P & ID of a Typical Petfood Dryer

a temperature detecting device. The product flow through the vessel is controlled by a variable speed positive displacement device (rotary valve, auger or metering roll) in order to maintain a constant velocity through the meter. The meter sends an analog signal proportional to the product moisture content to the Dryer Master.

The general form of the dynamic model used for the dryer is given by equation 5.1.

$$M_{out}^{P} = f(d, x, u, t) \tag{5.1}$$

where: d is a set of design variables

  x is a set of state variables

  u is a set of manipulated variables

  t is time

  M_{out}^{P} is the predicted outlet moisture of the petfood product

Since dryers tend to exhibit long process dead time and significant process response time it was necessary to add a dead time compensator as shown below, Lee et al (1990). Equation 5.2 is the predictor corrector term. This term is filtered by an exponential filter of the process time constant.

$$M_{t}^{corr} = e^{(\frac{-\Delta t}{\tau_{p}})} M_{t-\Delta t}^{corr} + (1 - e^{(\frac{-\Delta t}{\tau_{p}})}) M_{out_{t-\Delta t}}^{P} - M_{out_{t-\Delta t-d}}^{P} \tag{5.2}$$

A predicted trajectory of the actual controlled variable, M, is given by equation 5.3.

$$M_{t}^{control} = M_{t} + M_{t}^{corr} \tag{5.3}$$

where M_{t} is the actual measured moisture.

Equation 5.4 calculates the error term based on the desired setpoint, M^{*}, and the predicted process output based on previous control action.

$$\varepsilon_t = M^* - M_t^{control} \qquad (5.4)$$

Equation 5.5 calculates the change in the error term from the previous control interval.

$$\Delta\varepsilon = \varepsilon_t - \varepsilon_{t-\Delta t} \qquad (5.5)$$

The change in the temperature output, equation 5.6, is calculated by solving the GMC control equations using the model of dryer, equation 5.1.

$$\Delta T_t = \frac{1}{\beta} \left\{ \Delta M_{in} - \Delta M^{control} - \tau_p \left[K_1 \Delta\varepsilon + K_2 \varepsilon \Delta t \right] \right\} \qquad (5.6)$$

where β is a model parameter.

The first two terms of equation 5.6 result from the drying model, while the last two terms are derived from the GMC reference trajectory.

The control system is also capable of monitoring other variables in the dryer such as product temperatures during and after drying, conveyor belt(s) speed, product height on the conveyor belt and exit air humidities. All instrumentation is interfaced to the computer system using a modular I/O system. This allows the system to expand and include as much instrumentation as is necessary for further applications. Since specific pet food products have different bulk density specifications, each product requires its own moisture calibration curve. The calibrations are stored and updated on-line by the computer system whenever the operator provides reference moisture readings. The computer system software has been designed so that its functionality is controlled by a configuration file. This allows software features to be changed easily or customised for a particular site application.

The display and keypad are contained in a NEMA 4 enclosure which will allow installation of the Dryer Master display on the plant floor if required. A photograph of a typical system for a desk mount installation is shown in Figure 5.4. The operator interface would be the same for the plant floor installation. All measured parameters can be alarmed and defined as critical

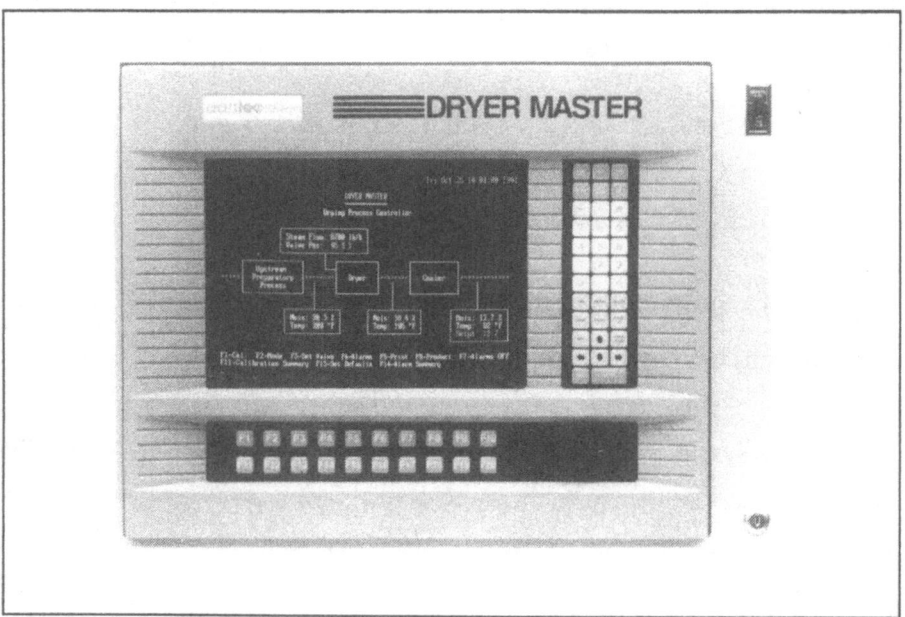

Figure 5.4 Dryer Master

to automatic control. This means that parameter values can cause alarms if out of a normal operating range but will not defeat automatic control unless the parameter has been defined as critical to automatic control or proper dryer operation.

The Dryer Master provides two options for obtaining a permanent record of the drying operation. A printer can be attached to the computer system to provide a semi-graphical record of dryer operation either continuously or the last 24 hours on demand. The second option allows communications directly with the Dryer Master computer using another computer. This will allow direct access to the dryer operating history for additional manipulation, or customised analysis and plotting.

In standard pet food applications, the Dryer Master will manipulate the temperature of the final drying zone to a specified target, and save energy due to the reduction of overdrying. The control system provides a real time display of all measured operating parameters. The printed copy of the dryer's operating history will allow quality control to pinpoint any product not meeting specifications and allow operating staff to document operations.

5.6 RESULTS

The operation of this dryer has been monitored continuously for a period of about one year. Because there was no fuel meter on the dryer, decreases in fuel consumption will be estimates only. The comparison between operating runs can be difficult since the ability to dry a given product can vary with factors such as ambient air temperature and relative humidity, efficiency of the air preheater and how the product is processed before the dryer. For the purpose of comparison between manual and automated control, two production periods for the same product will be compared. The product moisture and dryer temperatures are shown in Figure 5.5 for a manually controlled production period in June 1989. In July 1989, the same product was produced using automated dryer control. The product moisture and dryer temperatures are shown in Figure 5.6.

In comparing control performance, the mean and standard deviation of the product moisture will be considered. For the manually controlled production run, the moisture content averaged 9.5% w.b. with a standard deviation of 0.833. This product had a maximum allowable moisture content of 11.5% w.b. Due to a level of uncertainty in the dryer operation, an average safety margin of 2% w.b. moisture was maintained by the operator. For the automated production run, the average moisture content was the same as the operator specified setpoint of 10.0% w.b. The standard deviation was reduced to 0.306.

These two production runs also demonstrate the difficulty in comparing production runs on the basis of energy reduction. During the majority of the manual run, zones 1 and 2 air temperatures were set at 115°C and zone 3 at 125°C. For the automated run, zones 1 and 2 were set at 100°C and zone 3 averaged about 105°C. During the research phase it was found that a 15°C change in zone 3 alone accounted for a 1% change in product moisture. Therefore the full decrease in drying air temperature and the related energy savings cannot be attributed to the automated moisture control system.

Figure 5.7 shows normalised product moisture distributions for both the manual and automated runs. The manual controlled data shows the large safety margin of overdrying required to ensure that all the product will be below 11.5% moisture. The automated moisture control results in an average

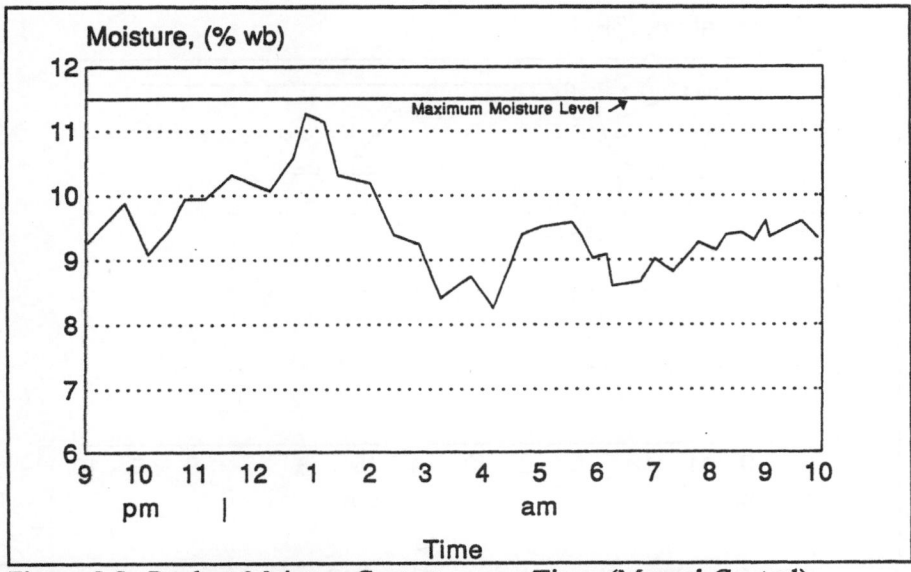

Figure 5.5a Product Moisture Content versus Time, (Manual Control)

moisture level of 10.0% are equal to the operator specified setpoint. With the improved standard deviation demonstrated here, management found that they could raise the moisture setpoint 10.5% and still be assured of all the product having a moisture below the maximum allowable level as shown in Figure 5.7.

With the demonstrated increase in throughput of 1%, (10.5 - 9.5), from an average decrease of 15°C in the operating temperature of zone 3 an estimated savings of 16.6% of the fuel for zone 3 may be realised. Since approximately one third of the total dryer fuel consumption is in zone 3, the total energy savings would be about 5.6%.

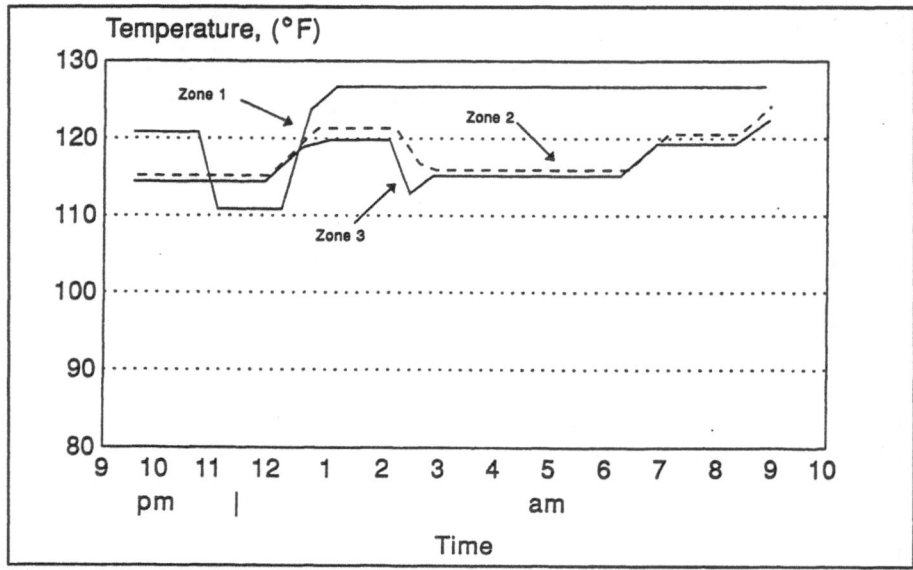

Figure 5.5b Zone Temperatures versus Time, (Manual Control)

5.7 COST BENEFIT ANALYSIS

The two easily identified economic benefits resulting from the reduction in overdrying are energy savings and yield improvements. The potential energy savings are 5.6% of the total dryer energy consumption resulting in a potential energy savings of about $CDN6,720/year.

The improvement in yield will have a significant benefit for the pet food producer. The potential of a 1% increase in average moisture will result in the equivalent increase in yield since the final product is sold by weight. The increased revenue possible is estimated to be $CDN78,000/year based on a yearly production of 12,000 tonnes at a value of $CDN650/tonne.

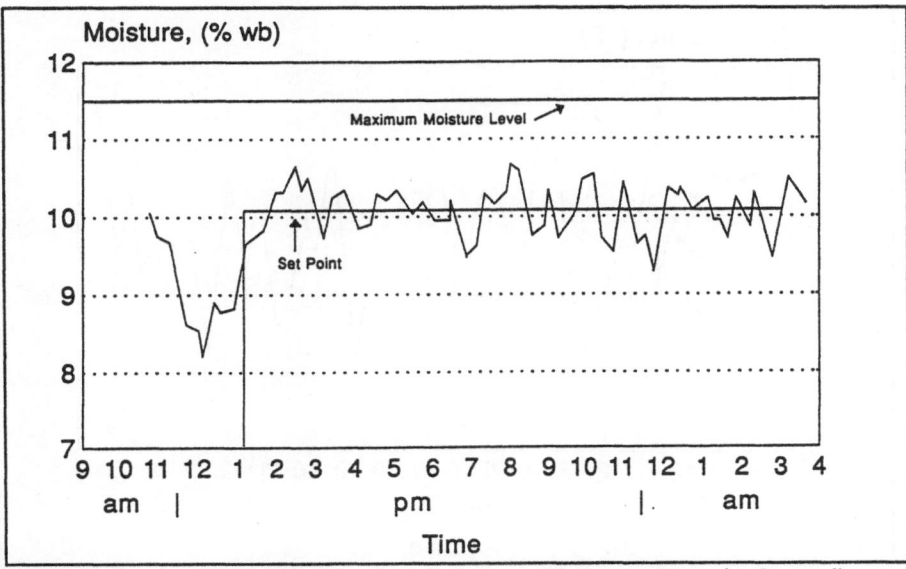

Figure 5.6a Product Moisture Content versus Time, (Automatic Control)

There are other benefits to installing a control system which are more difficult to estimate. The improved ability to monitor the process normally leads to improved quality through increased operator awareness. In this case, by continuously monitoring the product moisture levels, product that must be redried or returned for reworking can be virtually eliminated. These benefits are estimated at 0.5% of total yearly product value, $CDN39,000. Therefore the total economic benefit for the site considered in this study is estimated at $CDN123,720/year.

The estimated installed cost of the control system in this study is $CDN68,000. This results in a payback time of about 6 months.

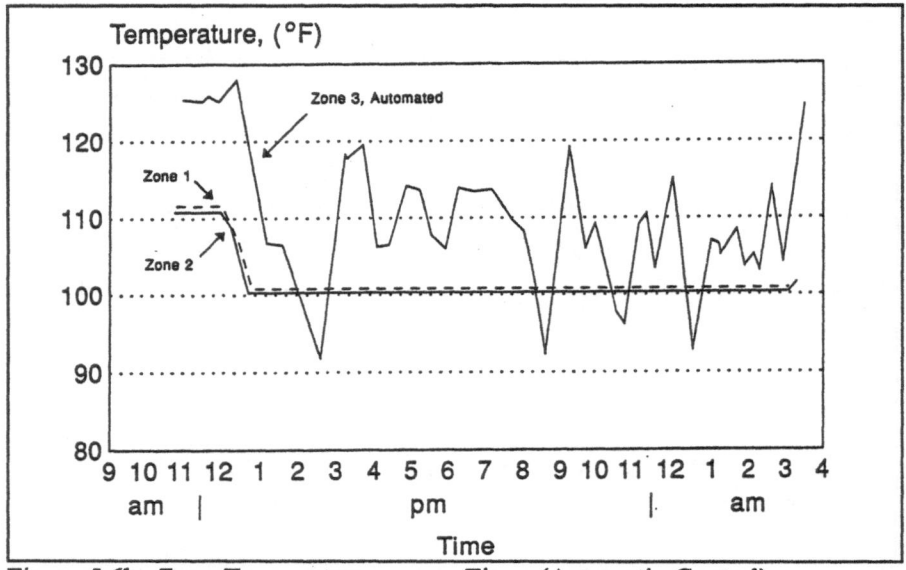

Figure 5.6b   Zone Temperatures versus Time, (Automatic Control)

5.8 DISCUSSION

The two major benefits of the pet food dryer control system are the energy savings and yield improvement. These benefits are achieved by reducing the standard deviation of the product moisture content and then being able to raise the average moisture content closer to the maximum allowable moisture content. The cost benefit analysis provided in the previous section was estimated for a particular drying system. The results of analysis for different drying systems will depend on production throughput, the number of drying zones and the overall dryer efficiency. The quality of manual dryer control before the implementation of an automated control system will also affect the magnitude of any benefit.

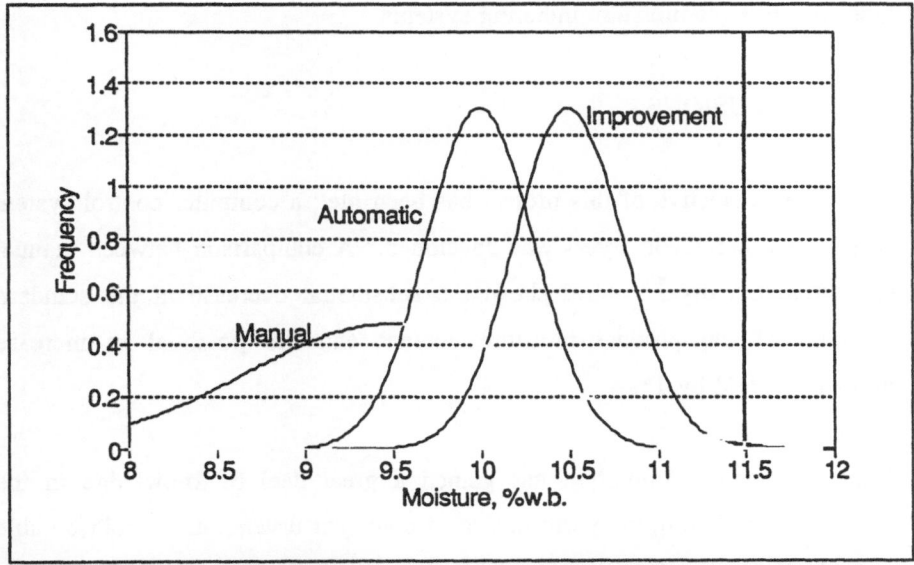

Figure 5.7 Normalised Moisture Distributions (Manual versus Automatic)

The on-line metering of the product moisture will require a period of custom calibration prior to installation. This work will be required in order to span the moisture meter electronics for the full range of moistures and product densities that will be encountered at a given site. During this phase the temperature correction factors can be found which will reduce the commissioning time required for moisture meter calibration. Since the moisture meter will be customised for each application, the accuracy of the meter and therefore the quality of control will be affected if the product density or temperature fall outside the defined range.

Because the control system can handle any number of products through specific calibration curves, and the process control functions can be customised, the dryer performance and product quality will be superior to

129

other less flexible moisture metering systems.

5.9 CONCLUSIONS

The overall objective of this project has been met; a computer control system for pet food conveyor dryers was developed. A comparison between manual and automated dryer control showed a substantial decrease in the standard deviation of the product moisture content and the potential to increase production yield by 1%.

Dantec Systems Corporation has gained a great deal of knowledge in the mechanisms affecting the performance of conveyor dryers, and should be able to apply the technology developed in this project to other conveyor drying systems. Considering the cost benefit analysis, the system should have a payback period of less than one year which would make the system an attractive addition to a conveyor drying application.

Although the economic incentives for controlling moisture are significant the non-linear nature of the process as well as the difficult moisture measurement problems result in a very difficult control problem.

5.10 ACKNOWLEDGEMENTS

We wish to acknowledge the support of the Ontario Government and the continued interest of Dantec Systems Corporation in this research area.

The technical assistance of Dr Gerald R. Sullivan in implementing the GMC model based control technology on a conveyor belt dryer was greatly appreciated.

5.11 REFERENCES

Bakker-Arkema F.W., Anderson J.C. and Eltigani A.Y. (1988) Drying Fuel Cost Control, Grain Age, March 4,8.

Behlen (1986) Owner's Manual, Model 850 Continuous Dryer.

Carr-Brion (1986) Moisture Sensors in Process Control, Elsevier Science Publishers, New Jersey, USA.

Dantec (1992) Product Literature, DRYER MASTER, Dantec Systems Corporation, Waterloo, Ontario.

LAW (1986) Dryer operations manual.

Lee P.L. and Sullivan G.R. (1988) Generic Model Control (GMC). Computers in Chem. Eng. 12(6):573-580.

Lee P.L., Sullivan G.R. and Zhou W. (1990) A New Multivariable Deadtime Control Algorithm. Chem.Eng.Commun. 91:49-63.

Moreira R.G. and Bakker-Arkema F.W. (1990) Journal of Agricultural Engineering Research. 45:107-110.

NIR Systems (1989) Product Literature - 550 Spectrophotometer, NIR Systems Inc., Silver Spring, MD, USA.

Reilly P.M., Sullivan G.R., Whaley M.G. and Fleming J.F. (1988) Application of EVM for On-line Moisture Measurement in Drying Processes, Proceedings of Sixth International Drying Symposia, Versailles, France.

CHAPTER 6

ECCENTRICITY CONTROL OF A CABLE JACKETING LINE

6.1 INTRODUCTION

In the mid 1980's, an automatic jacket thickness and eccentricity control system was installed on a number of polyethylene telephone cable jacketing lines at Northern Telecom (Canada) Kingston Works Plant. At the time, it was claimed by the supplier that by switching from a manual to an automatic operation, material savings of 4 to 8 % could be achieved (Brunner and Merki, 1985). A 10 % material savings on the jacket material provides a one year payback on the installed cost of the control system.

The value of a jacket control system when performing as designed is widely acknowledged within the industry (Boggs et al, 1983). However, for a number of reasons the Kingston system has not performed "as designed" on a reliable basis. The controllers were tuned for very sluggish response due to perceived stability problems at the time of commissioning. Controller response time is of the order of 5 to 10 min. This is considered acceptable under the majority of conditions, where the goal is to prevent long-term drift in the process. However, the presence of short-term disturbances has necessitated a thicker than optimal wall thickness to compensate for non-zero eccentricity. This approach negates any direct economic savings due to jacket material savings.

This is not to say the system is not considered to be of value. The current benefit is perceived as being qualitative in nature given that an operator can leave the line unattended for a period of time without having to worry about significant drift in the process. However, this paper describes the preliminary results of a project that set out to determine the achievable quantitative performance given a properly tuned multivariable controller. The characteristics of the process were such that a Generic Model Controller (GMC) was well suited as the candidate multivariable control algorithm.

6.2 THE APPLICATION

The process involves passing a multipair conductor telephone cable core through an extruder where a cylindrical tube of polyethylene is applied. The performance goal is to maintain a uniform jacket of a specified thickness around the cable and along the cable length. The overall process is illustrated in Figure 6.1. Though the total length of the line is 75 m (250 ft), the suspended length from the extruder to the end of the cooling trough is less than 15 m (50 ft). Eccentricity is the difference in opposite wall thicknesses as is shown in Figure 6.2. Thus, the performance goal can be restated as the need to maintain zero eccentricity.

Jacket thickness is controlled by manipulating the speed of the extruder.

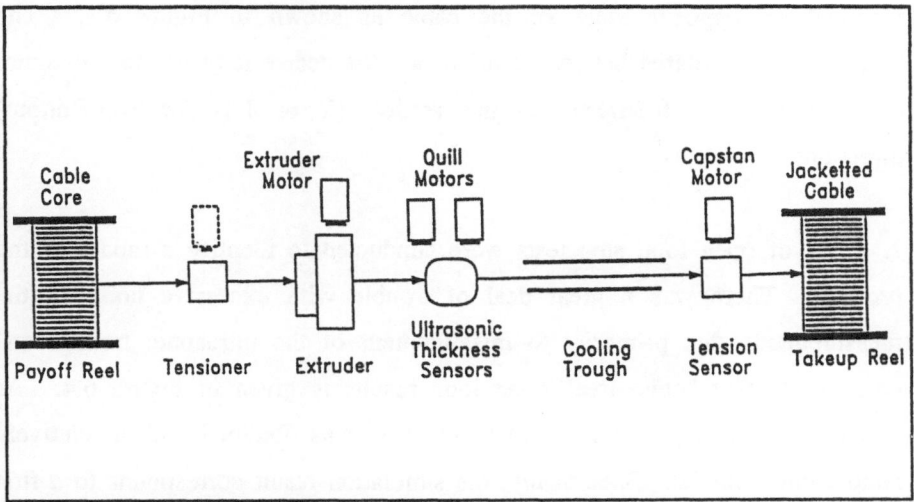

Figure 6.1 Illustration of the cable jacketing line

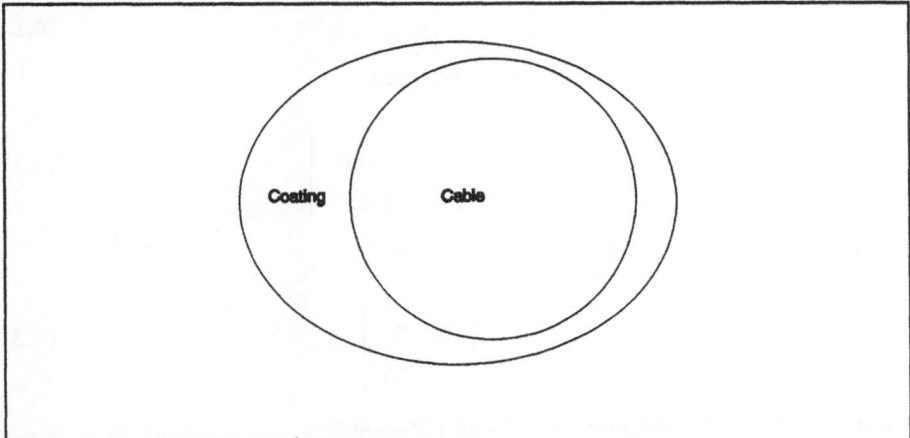

Figure 6.2 Eccentricity of Cable Coating

Eccentricity is controlled by horizontal and vertical positioning of a movable die (referred to as the quill) in the extruder head. The thickness (and indirectly the eccentricity) is measured with a pair of ultrasonic transducers mounted on opposite sides of the cable as shown in Figure 6.3. The transducers are rotated between ±90° to get the eccentricity in the horizontal (X or L-R for Left-Right) and the vertical (Y or T-B for Top-Bottom) directions.

A series of open loop step tests were conducted to identify a model of the process. There was a great deal of trouble with excessive noise in the measurements due primarily to misalignment of the ultrasonic transducers. One of the few "noise-free" open loop results is given in Figure 6.4. As indicated in the figure a good fit to the data was obtained with a relatively simple linear model. Specifically, the simulation result corresponds to a first order 2-input 2-output linear model:

$$dy/dt = \tau^{-1}(G\ u - y) \tag{6.1}$$

where the steady state gain matrix is given by:

$$G = \begin{bmatrix} -0.011 & 0.009 \\ 0.009 & -0.028 \end{bmatrix} \tag{6.2}$$

and the matrix of (dominant) time constants is:

$$\tau = \begin{bmatrix} 30 & 0 \\ 0 & 30 \end{bmatrix} \tag{6.3}$$

and the controlled and manipulated variables are:

u_1 = change in X quill position, %

u_2 = change in Y quill position, %

y_1 = change in X (Left-Right) eccentricity, mm

y_2 = change in Y (Top-Bottom) eccentricity, mm

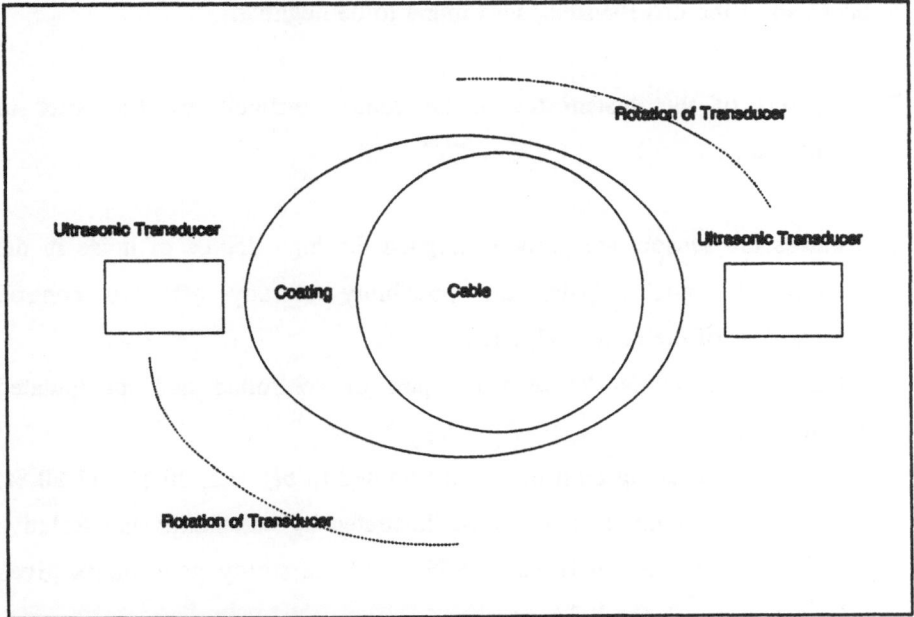

Figure 6.3 Physical Location of Transducers

In reference to Figure 6.4, the fit is good up to the 67 min mark, at which point a 0.5 mm offset in the Y-eccentricity develops. At this point there was a 50 % step reduction in the line speed that effectively results in a "drooping" of the cable in the vertical plane. The horizontal or X-eccentricity is seen to be unaffected. This is considered to be the major disturbance and as such was used to benchmark controller performance.

The process is predominantly a mechanical one with the positioning system driven by stepper motors. The first order response is attributed to the dynamic characteristic of the motors, together with the lag between the movement of the quill and the movement of the polyethylene. Deadtime (due to backlash in the drive system) was found to be neglible.

The aspects of the system that make control difficult are (in order of decreasing importance):

1) unless the sensors are properly aligned the high degree of noise in the measured signal negates the possibility of any effective control, regardless of the control algorithm in use,

2) there is interaction between the pair of controlled and manipulated variables, and

3) practically speaking quill moves are limited to between 20 % and 80 %. The degree of interaction is best illustrated by calculating the Relative Gain Array (RGA) of Bristol (1966). For the steady gain matrix given by equation 6.2, the RGA is:

$$RGA = \begin{bmatrix} 1.36 & -0.36 \\ -0.36 & 1.36 \end{bmatrix} \qquad (6.4)$$

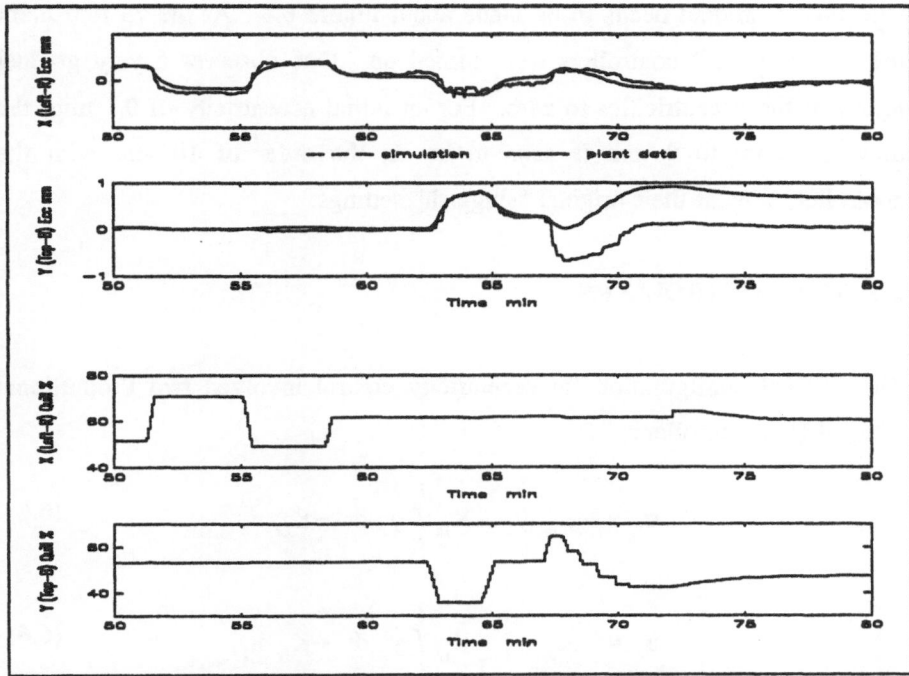

Figure 6.4   Open loop test result. Note 1) offset in Y-ecc after 67 min due
to speed change, and 2) controllers turned on at 73 min mark

One notes that the diagonal elements match the natural pairing for the
controllers, namely u_1 to y_1 and u_2 to y_2. The non-zero off-diagonal element
(-0.36) quantifies the magnitude and direction of the coupling.   The
observation is that moving the X-quill affects both the X and Y-eccentricities,
and moving the Y-quill affects both the X and Y-eccentricities. Thus, for
optimal performance one would expect to require a formal multivariable
controller that coordinates the moves of the X and Y-quills. The controller
designs considered, together with summarized performance results, will be
given in the next section.

139

One final comment needs to be made about Figure 6.4. At the 73 min mark the X and Y-quill controllers were turned on. One observes a very gradual return of the eccentricities to zero. For an initial eccentricity of 0.1 mm, the time to return to 0 mm is seen to be on the order of 10 min with the controller gains at their original "sluggish" settings.

6.3 THE CONTROLLERS

The original configuration for eccentricity control involved two Proportional Integral (PI) controllers:

$$u_1 = -K_{px} \, y_1 \, -K_{ix} \int_0^{t_k} y_1 \, dt \, +K_{fx} \, u_2 \qquad (6.5)$$

$$u_2 = -K_{py} \, y_2 \, -K_{iy} \int_0^{t_k} y_2 \, dt \, +K_{fy} \, u_1 \qquad (6.6)$$

where K_{px} and K_{py} are the proportional gains and K_{ix} and K_{iy} are the integral gains. The feedforward (FF) gains K_{fx} and K_{fy} were not originally implemented by the plant. However, given the results of the RGA analysis, some form of decoupling compensation appears to be called for.

The closed loop models formed by the above PI controllers in combination with the process models of equation 6.1 are second order. On this basis, the gains can be calculated to provide a specified overshoot and settling time in response to a change in setpoint for the individual loops. Simulation tests indicated that a specification for a 20 % overshoot with a 60 sec settling time provided adequate regulatory action, given the restriction of keeping quill

140

moves between 20 and 80 % for the benchmark disturbance. This required that $K_{px} = 68$, $K_{py} = 28$, $K_{ix} = 15$ and $K_{iy} = 6$. For ideal feedforward action, the proportional gains combine with the steady-steady process gains to give $K_{fx} = 0.8$ and $K_{fy} = 0.3$.

As a formal multivariable algorithm, a Generic Model Controller (GMC) was adopted. Details of the GMC approach can be found in Lee and Sullivan (1988). For a linear first order model of the form given by equation 6.1, the GMC control law is:

$$u = G^{-1}\left(\tau\left[-K_1 y - K_2 \int_0^t y \ dt\right] + y\right) + u_o \tag{6.7}$$

where u_o gives the initial positions of the quill. By comparison with equations 6.5 and 6.6 one notes that GMC is essentially a PI controller with formal decoupling provided by the inverse of the steady-state gain matrix. In a fashion similar to that adopted for the PI design, the gains K_1 and K_2 are selected to provide a desired trajectory in response to a setpoint change. In this application it was found the GMC gains were best taken as:

$$K_1 = 1/t_r \tag{6.8}$$

$$K_2 = (1/t_r)^2 \tag{6.9}$$

For servo action, this corresponds to an underdamped response with a 20% overshoot and t_r as the rise time. This design trajectory was found to provide adequate integral action for the regulation problem encountered in this application. In addition, it matched the design approach used for the PI controller and consequently would provide a basis for fair comparison. Simulation tests indicated that the minimum rise time was 10 sec if quill moves were to be restricted to between 20 and 80 % for the benchmark disturbance. Thus, the adopted GMC gains were K_1 = diag[0.1 0.1] and K_2 = diag[0.01 0.01].

Hardware Details

The quill positions are measured by LVDT's mounted on the quill itself. The wall thicknesses are measured by 5 MHz spherical ultrasonic tranducers mounted on a rotating rocking head placed in a water bath. The controller is a Texas Instruments (TI) 560 PLC operating with a 1 sec sampling time. The TI PLC is capable of controlling 64 PID loops with full programming capability in a BASIC-like language. Data acquisition is performed by a microVAX 4000 operating in a supervisory role and operating with a 5 sec sampling time.

6.4 RESULTS

A series of simulation tests were conducted to assist in the design of the PI and GMC controllers, as well as to determine their relative performance prior to actual implementation. Given experience with the open loop field tests, it was felt that the benchmark disturbance would be the equivalent of +0.5 and -0.5 mm near steps in the Y-eccentricity at 5 min intervals.

142

Performance Goal

The performance goal is benchmarked against 25 mm nominal diameter cable with a wall thickness of 1.4 mm. The acceptable tolerance for eccentricity (thickness) is ±0.05 mm (±3.5 %). The sensors are technically capable of measuring to ±0.003 mm (±0.1 thou). In open loop, the observed "normal" noise was ±0.01 mm (see Figure 6.4).

The major disturbance is a ±50 % near step change in the line speed that occurs whenever the operator has to work on the line as it runs (for example, a tape changeover and weld operation). Minor disturbances include: 1) X-Y movement of the cable due to changes in tension, 2) X-Y movement of the sensor heads as they are continuously rotated between ±90°, and 3) minor fluctuations in cable speed. It was felt that regulation on the order of ±0.02 mm (≈±1 thou) was achievable and this was set as the performance goal.

Simulation Results

The nature of the benchmark disturbance, together with individual results for a PI and GMC test, is indicated in Figures 6.5 and 6.6. The horizontal lines on the eccentricity plots correspond to the ±0.02 mm performance goal. In both figures, the "blips" in the Y-eccentricity correspond to the timing of the disturbance. In Figure 6.5, one notes that although the quill moves stay within the 20 and 80 % guideline, both the X and Y-eccentricity exceed the ±0.02 mm goal. On the other hand, response is quite rapid with recovery in under 1 min. Figure 6.6 illustrates that the GMC response is even more rapid. In addition, effective decoupling keeps the X-eccentricity within the ±0.02 mm goal.

Table 6.1 summarizes the simulation results for the PI (K_f gains zero), PI+FF (K_f gains non-zero) and GMC designs. The Integrated Absolute Errors (IAE) of the X and Y eccentricities are given as quantitative performance measures for a 20 min test. Each design was tested against a "good" and "poor" model. For the good model, the process gains were those given by equation 6.2 and match those used in the actual design of the controllers.

For the poor model, the process gains were given as:

$$G = \begin{bmatrix} -0.011 & 0.009 \\ \underline{0.004} & \underline{-0.053} \end{bmatrix} \qquad (6.10)$$

with the altered gains underlined. The performance with the poor model indicates the degree of robustness of the controller designs. It was noted that in certain open loop field tests the process gains for the Y-eccentricity could range by the amount given by equation 6.10.

The second last column of Table 6.1 gives the total IAE score normalized to the "worst" PI performance. The last column states whether the X (L-R) eccentricity was maintained between the ±0.02 mm goal. One notes that PI performance actually improves with the poor model, mainly because the effect of the gains given by equation 6.10 is to reduce the degree of interaction. For this same reason, the addition of feedforward action with the PI+FF controller does not significantly improve overall performance. By

144

contrast, the GMC performance is consistently better by a factor of ≈2.5 (0.28 versus 0.69).

Table 6.1. Summary of simulation results.

Control	Model	IAEx	IAEy	IAEx+y	Normal	L-R OK
PI	good	9.4	21.1	30.5	1.00	no
	poor	3.8	17.1	21.0	0.69	yes
PI+FF	good	5.0	19.2	24.3	0.80	yes
	poor	5.6	17.1	22.8	0.75	no
GMC	good	0.0	8.3	8.3	0.27	yes
	poor	1.8	6.7	8.5	0.28	yes

6.5 CONCLUSIONS

The simulation results indicated that both the PI+FF and GMC designs could provide significantly improved performance over the existing (badly tuned) PI controller. The process model used to properly design the PI+FF controller was also required as the basis for the GMC algorithm. Thus, the design effort for the controllers was judged to be the same. On the other hand, it appeared that for the benchmark disturbance, GMC could provide a factor of two improvement in regulation. On this basis, the GMC design was implemented on the line with the least noise. Figure 6.7 gives a GMC closed loop test result with a speed change.

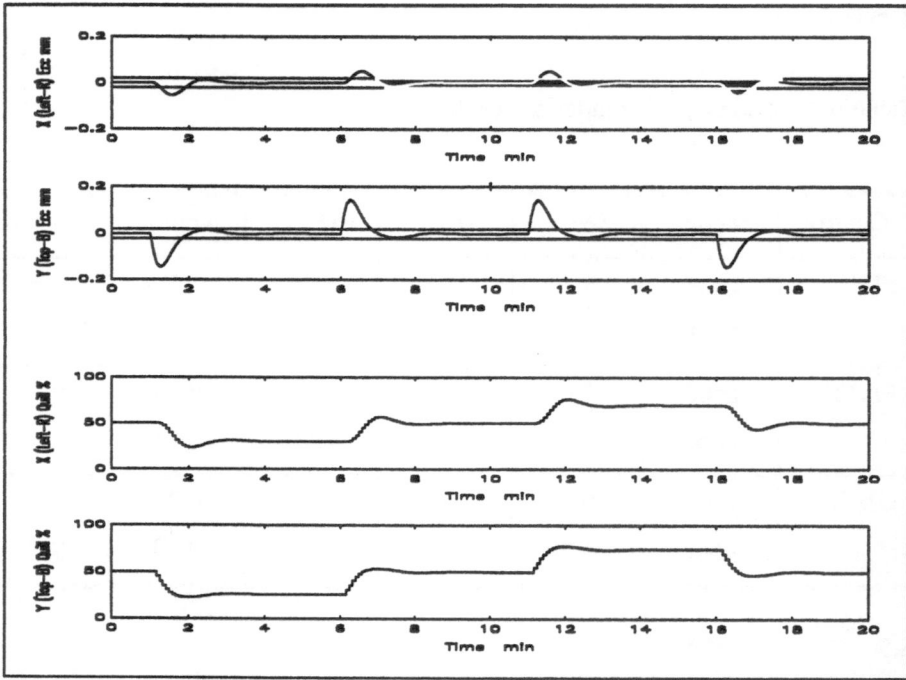

Figure 6.5   PI simulation result with "good" design.   Note 20 to 80% quill range, and both X and Y-ecc exceed +0.02 mm goal.

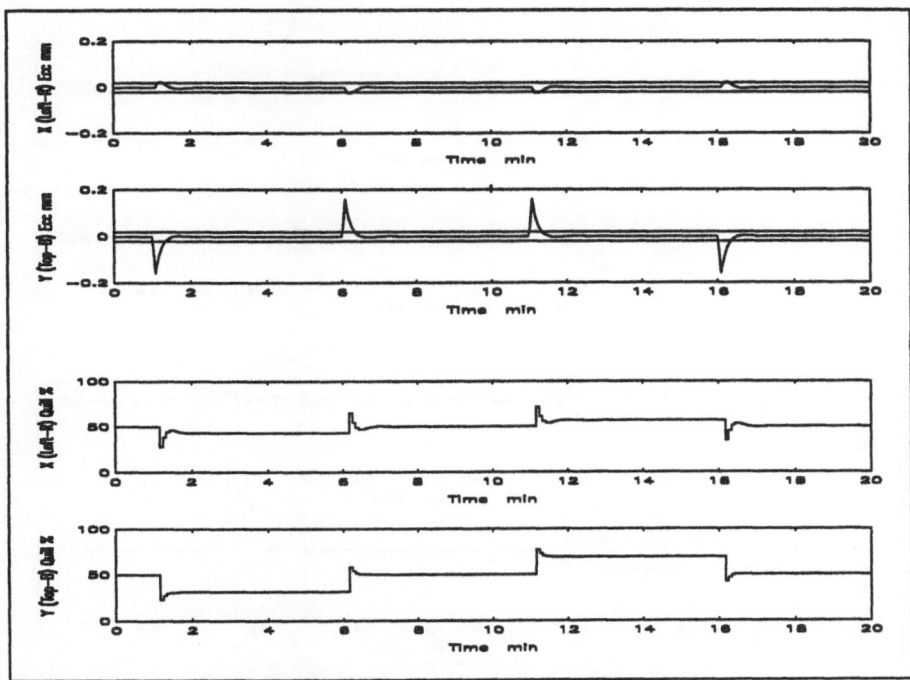

Figure 6.6   GMC simulation result with "poor" design.   Note 20 to 80%
quill range, and only Y-ecc exceeds +0.02 mm goal.

The X-eccentricity is maintained within the ±0.02 mm goal and the
Y-eccentricity briefly exceeds the goal at the time of the speed change.   On
this basis GMC is judged to be successful.   Further field trials are being
conducted before a decision is taken to reduce the jacket thickness setpoint.
This has to be coordinated with a revised speed control scheme to mitigate
the severity of the Y-eccentricity disturbance.   The overall result should be
an improvement in the performance of the eccentricity control system, in
terms of both its reliability and its ability to deliver the minimum optimal
jacket thickness.

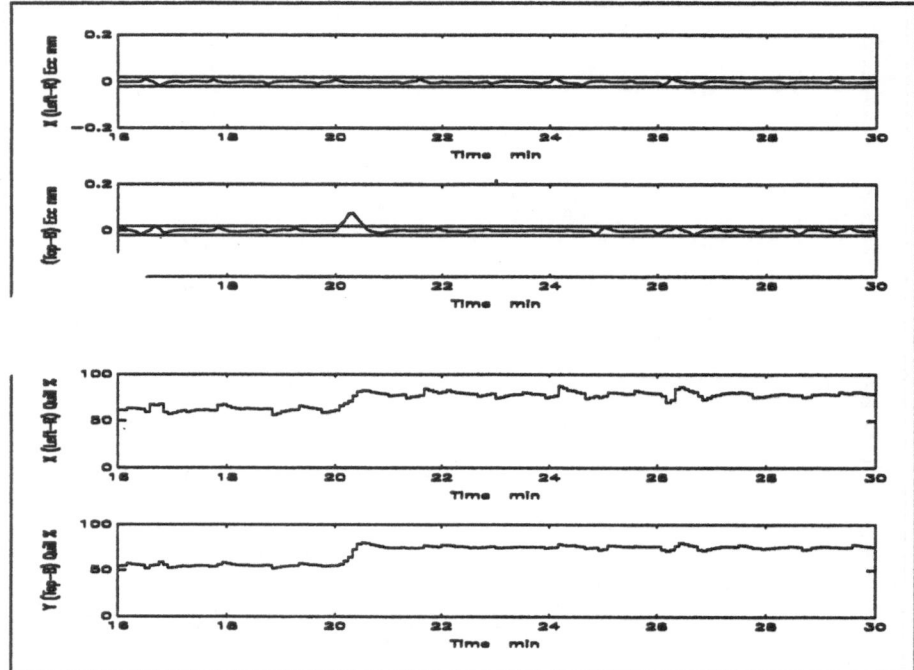

Figure 6.7   GMC closed loop plant test result with speed change at 20 min.
Note only Y-ecc exceeds +0.02 due to speed change.

6.6 ACKNOWLEDGEMENTS

This work was supported by a Grant from Northern Telecom Canada
Limited.   The assistance of Joe Oresti and personnel from the
Communication Cable Division (Kingston Works) is gratefully acknowledged.

The technical assistance of Reza Neshat during his work term at Queen's
University is also acknowledged.

6.7 REFERENCES

Boggs L.M., Isley A.M. and Levenhood J.W. (1983)   Ultrasonic Jacket Thickness and Eccentricity Monitor and Control System.   Proceedings 32nd International Wire and Cable Symposium, Atlanta 359-362

Bristol E.H. (1966)   On a New Measure of Interaction for Multivariable Process Control. IEEE Trans. Auto Control. Vol AC-11, 1:133-134

Brunner M. and Merki H.A. (1985)   On-Line Monitoring and Control of Jacket Thickness and of Eccentricity.   Zumbach Electronics Corp., Mt. Kisco, NY.

Lee P.L. and Sullivan G.R. (1988) Generic Model Control (GMC).   Comp. and Chem. Eng. 12, 6:573-580

CHAPTER 7

THE DEVELOPMENT OF A NONLINEAR ADAPTIVE GENERIC MODEL CONTROLLER FOR CHEMICAL REACTION QUALITY CONTROL

7.1 ABSTRACT

The purpose of this paper is to outline the evolution of a model-based control strategy to control the output quality variable of a chemical reaction process. Achieving high performance control of this quality variable was a challenging task, as the process has many characteristics that make traditional control techniques difficult to apply successfully. These characteristics include severe nonlinearities, slow and time - variant process dynamics, and infrequent and noisy process measurements. In the end, the loop was successfully closed by incorporating a nonlinear process model into a modified Generic Model Control (GMC) framework. An adaptive algorithm was used to update a single parameter in the nonlinear model in order to ensure model prediction accuracy. Although the control strategy required approximately 2 man-years of effort before it performed up to our expectations, the strategy now yields economic benefits in the range of $CDN 2,000,000/year by reducing off-spec product.

7.2 INTRODUCTION

The use of computer-based control for chemical processes has rapidly expanded in the last ten years. The use of linear model-based controllers such as Dynamic Matrix Control and Internal Model Control has permitted the control engineer to solve multivariable control problems that would have been impossible to solve in the early 1980's. Further extensions of these technologies now allows the incorporation of process constraints and optimization concerns into the controller design, giving the control engineer much more flexibility.

Yet, there are still many processes for which high performance process control is difficult. These are processes with strong input-output nonlinearities, and complex, time-variant process dynamics. In chemical systems, these characteristics are usually found in and around the reaction systems.

Until recently, we have not had the control technology available to deal with these problems. However with the development of practical nonlinear control algorithms in the late 1980's, we can now begin to address these problems.

This paper describes the development of one such control strategy, which uses the nonlinear form of Generic Model Control algorithm as its basis. Our reason for attempting to close this loop is purely an economic one: our customers are demanding products with smaller and smaller variances in quality so that they may optimize their production systems. In addition, to

Shell Canada, any reductions in off-spec production translates into increased production capacity.

There are many ways in which the development of the control strategy could be presented. I have chosen a chronological approach because I feel it best describes the challenges and hurdles that must be overcome in successfully closing such a loop.

7.3 CHRONOLOGICAL DEVELOPMENT OF THE CONTROL STRATEGY

7.3.1 July 1990 - The Process

Due to the proprietary nature of the process under examination, we cannot discuss the process in detail, other than to say that we are concerned with controlling a product quality variable resulting from a chemical reaction process.

The process has several characteristics which make high performance control difficult.

1)   The process is highly nonlinear with respect to the manipulated and disturbance variables.

2)   The process dynamics between the manipulated and controlled variables are both long and complex. There exists approximately two hours of deadtime and a dominant time constant of about 4

hours in the process. In addition, the dynamics of the process vary with the throughput of the plant, and the throughput can change frequently.

3)   The measurement of the controlled variable is infrequent, only once an hour, due to the testing procedure. In addition, the measurement is subject to a fairly high level of noise about 10% of its mean value.

The only positive process characteristic is that because there is only one controlled variable and one manipulated variable, the system can be designed as a single input / single output (SISO) system.

The process performance of the quality variable has traditionally been reviewed on an informal basis, but increasing customer specifications in 1989 and early 1990 forced a formal performance review. At this time, the quality variable was controlled by operator intervention assisted by a manual Statistical Process Control (SPC) system.

Evaluation of the process performance during this period clearly demonstrated that the process was not performing up to our expectations. Figures 7.1 and 7.2 present time series and autocorrelation plots of the typical process performance. The autocorrelation of the time series is presented so the controller performance monitoring techniques of Harris (1989) can be used. Harris indicates that a process that is achieving minimum variance control will not have any significant autocorrelations past the number of lags equal to the deadtime of the process. Given that the process has two hours

154

of deadtime, or two lags, it can be seen that the process is nowhere near minimum variance control, and there is much room for improvement.

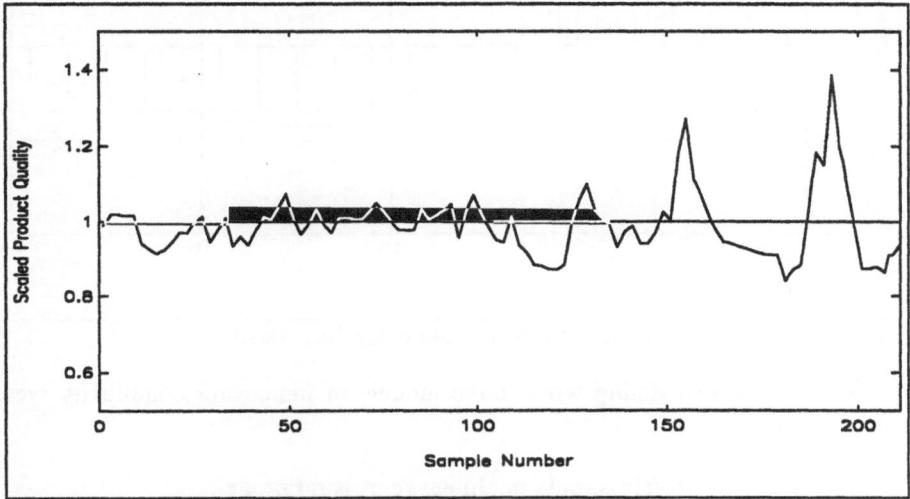

Figure 7.1 Product Quality Time Series for July 1990

7.3.2 August 1990 - The Model

As with all process control, good control performance begins with a good process model to describe the key process relationships.   Due to the importance of the quality variable on plant operation, model development was a joint effort between the Process Control group and the Process Engineering group. This made the introduction of the controller into the process operation much easier as we now had two groups supporting its implementation.

155

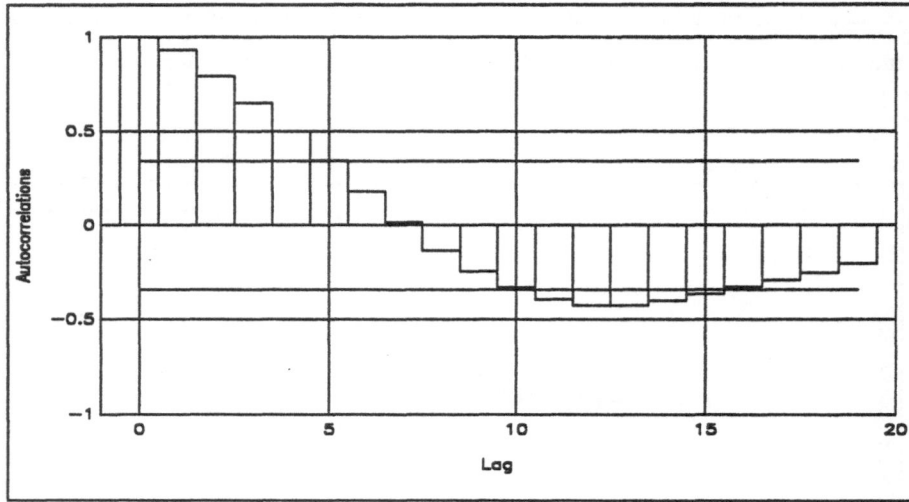

Figure 7.2 Product Quality Autocorrelation for July 1990

After a survey of existing work, three models of increasing complexity were available for use:

1)   A relatively simple nonlinear regression model;

2)   A model based on the simplification of a complex differential equation model;

3)   A complex differential equation model based on detailed mass balances around the reaction system.

After a detailed review of each model, we finally selected the second model. Our reasoning was as follows: we wanted the model to be based on theory, which eliminated the first model, while at the same time, we wanted a model based on a minimum number of inputs. The third model would require the use of many process inputs whose instrumentation was considered as secondary instrumentation, and not up to the same standards as other key process variables.

Our intention was to have the process model run on-line as frequently as possible, using the infrequent, off-line data to check the validity of the model. From our experience, we knew that the process was subjected to many and frequent unmeasured disturbances. Therefore, a key question was how were we going to handle the process/model mismatch. Do we use an adaptive method in order to update an appropriate model parameter, or do we let the controller contain the mismatch error?

In the end, we decided on an adaptive approach. Our main reason was that we had to sell the effectiveness of the model to our operations staff, so we had to have our model match the process output. One specific model parameter of model #2 was chosen to be updated. Initially, the values of all other model parameters were set from literature values.

The adaptation algorithm used was exponential data weighting with a variable forgetting factor (Goodwin and Sin (1984)). This technique was chosen due to the previous success of the technique in real-world adaptive control problems (Dumont (1982)). For the SISO case, implementation of the adaptation algorithm was straightforward. Unfortunately, there was still much trial and error tuning of the adaptation algorithm to be done, as will be presented later.

The adaptation of the model occurs by having the operations staff enter the off-line quality measurement into a special program on the computer system,

157

which retrieves the model output for the same time period and runs the adaptation algorithm. Clearly, the success of the controller is based on the timely and frequent entry of the off-line measurement. Fortunately, the operations staff quickly realized this fact, and the off-line entry had not been a problem.

7.3.3 October 1990 - The Controller

The implementation of the control strategy was aided by the fact that the process model obtained using model #2 could be partitioned into two sets of equations. The first and very small set describes the effect of the manipulated variable on the controlled variable. The standard Generic Model Control algorithm could be directly applied to this set of equations. The second set of equations was only required for the off-line update algorithm. They generate the model output at the same point where the off-line quality measurement is made. The entire system, the model, the controller, and the adaptation algorithm, were all implemented in FORTRAN. No special numerical routines were required, other than a 4th-order Runge-Kutta method to solve the ordinary differential equations present in the full model.

Given the high performance requirements of the controller, the GMC tuning constants were set very aggressively with an ξ of 3.0 and a τ of 2.0 time units. The loop was first closed in October 1990. Figures 7.3 and 7.4 give the initial performance of the controller.

As it can be seen, the initial performance of the controller has improved over manual operation, as the overall variance of the quality has been reduced. This can be seen by comparing Figure 7.4 to Figure 7.2. Note, though, that the autocorrelations are still significant after lag 2, indicating that we have not yet achieved minimum variance control.

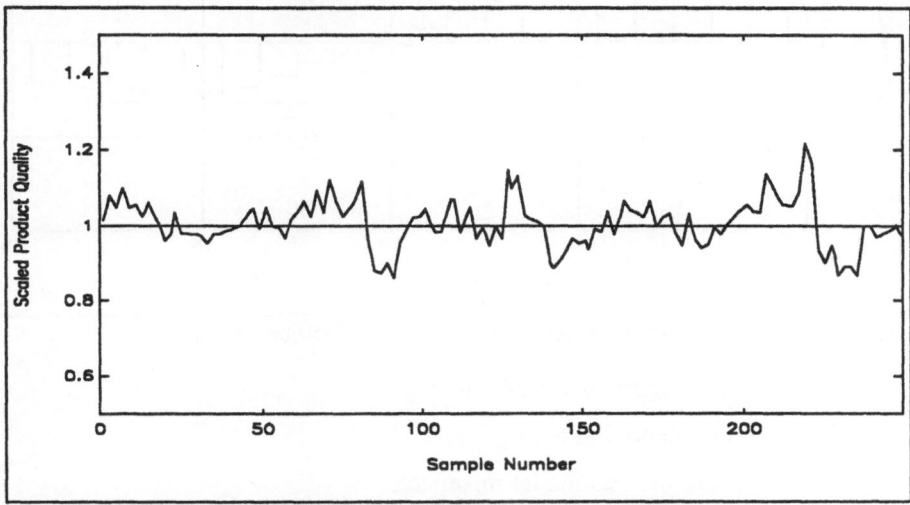

Figure 7.3 Product Quality Time Series for October 1990

7.3.4 November 1990 - Tuning of the GMC Controller

Unfortunately, less than three weeks later, the following performance as shown in Figures 7.5 and 7.6 was observed with the same controller.

Note that the autocorrelations for November do not die out very quickly. This indicates that the process has some very slow dynamics which are dominating the overall performance of the control strategy. Several reasons for this poor performance were hypothesized:

159

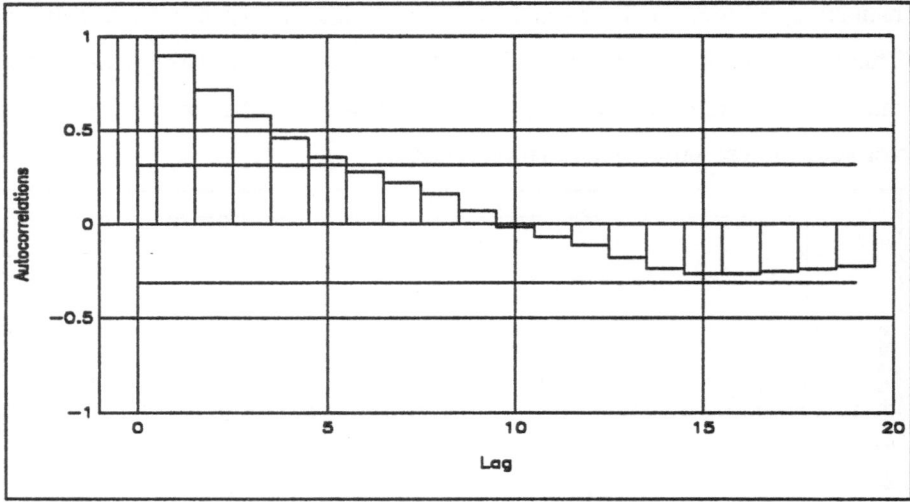

Figure 7.4   Product Quality Autocorrelation for October 1990

1)   overly-aggressive GMC tuning

2)   poor adaptation tuning

3)   severe process/model mismatch

It took us a great deal of time to resolve all these issues. We first addressed the easiest topic: GMC tuning. The tuning of the GMC controller was found to be very aggressive, given the prospect of plant/model mismatch. The controller as initially tuned often created a severe overshoot of the quality variable to get onto setpoint quickly. The problem with this approach is that the quality variable really does not blend; a blend of material above setpoint with material below does not have the same quality as material produced at setpoint.

To this end, we revamped the structure of our GMC controller. Firstly,

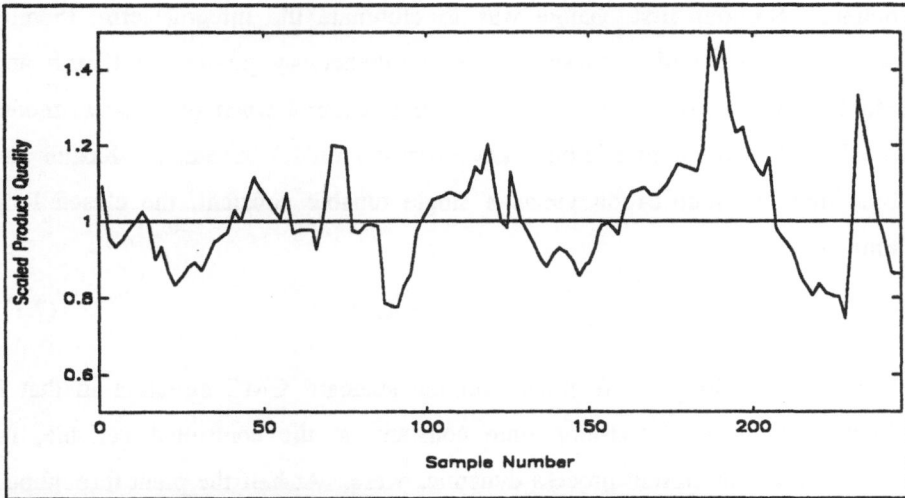

Figure 7.5 Product Quality Time Series for November 1990

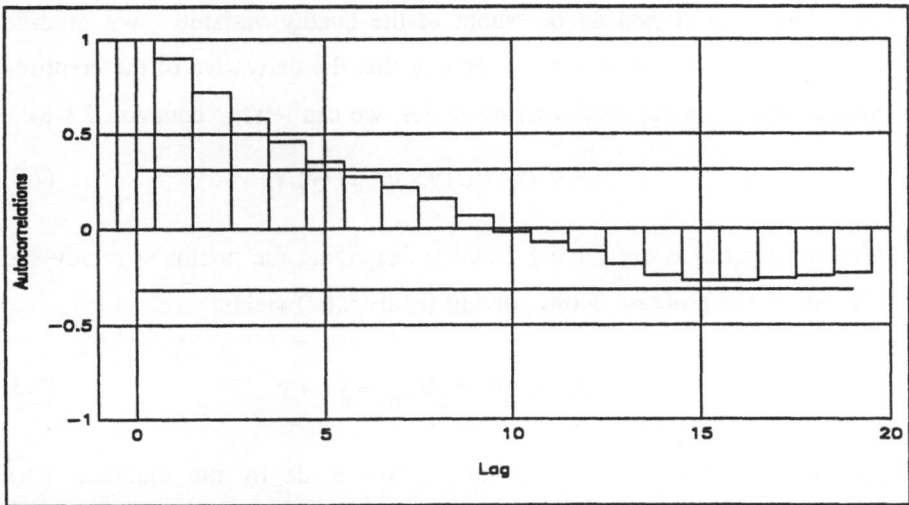

Figure 7.6 Product Quality Autocorrelation for November 1990

examination of the control moves indicated that the integral term of the GMC
control algorithm was not contributing anything significant to the controller

161

output.   So, our first change was to eliminate the integral term, thereby eliminating one tuning constant.   As simultaneously proven by Kozub and MacGregor (1990), the integral term is not required when the process model is adapted in order to eliminate any process / model mismatch.   Kozub and MacGregor's modification yields a single tunable constant, the closed loop time constant:

$$(y_{sp} - y)/\tau_c = dy/dt \qquad (7.1)$$

Secondly, we found difficulties with the standard GMC structure in that it forced the same closed-loop time constant on the controlled variable, no matter what the current process dynamics were.   At half the plant throughput, the control actions doubled to keep the dynamics at the same closed loop rate.   This also caused an overshoot of the quality variable.   We modified equation 7.1 to overcome this.   Assuming that the derivative of the controlled variable with time has first-order dynamics, we can rewrite equation 7.1 as:

$$(y_{sp} - y)/\tau_c = (f(y,x,u,t,d) - y)/\tau_p \qquad (7.2)$$

where f(y,x,u,t,d) is the forcing function describing the nonlinear steady state behavior of the process.   Now, solving for f(y,x,t,d) yields:

$$f(y,x,u,t,d) = \frac{\tau_p}{\tau_c}(y_{sp} - y) + y \qquad (7.3)$$

Equation 7.3 shows that the control moves made by the modified GMC controller are dependent on the ratio of τ_p to τ_c.   We chose to replace the closed-loop time constant, τ_c, in equation 7.3 with:

$$\tau_c = \alpha * \tau_p \qquad (7.4)$$

162

Here, α is a tuning constant relating the closed-loop time constant to the open-loop time constant.   By setting α equal to one, the steady-state controller move to make the quality variable equal to its setpoint is enforced at all times. We have found this modification and an α value equal to 1 has steadied the control dramatically by minimizing the controller's desire to overshoot the final setpoint in order to get the process reacting faster.

7.3.5 June 1991 - Tuning of the Adaptation Algorithm

The GMC algorithm changes were implemented at the end of November, 1990, yet we still had occasional problems with the quality variable cycling with a frequency of about 24 hours. This was not resolved until June 1991 when we first used power spectrum techniques to analyze the controller performance. Figures 7.7 through 7.8 present the time series, autocorrelation and power spectrum for the June 1991 dataset.

Note the power spectrum shows that the quality variable is only really affected at cycle times greater than 5 hours. The then-current tuning of the adaptation algorithm assumed much higher frequency disturbances in the quality variable. The minimum exponential filter initially used was 2 hours. Therefore, the adaptation algorithm was permitting high frequency noise (between 2 and 5 hours per cycle) to pass into the updated parameter and therefore into the manipulated variable moves. This, coupled with the known amount of process model/mismatch, caused the plant to cycle. Retuning the adaptation algorithm so that the minimum exponential filter used was 4 hours

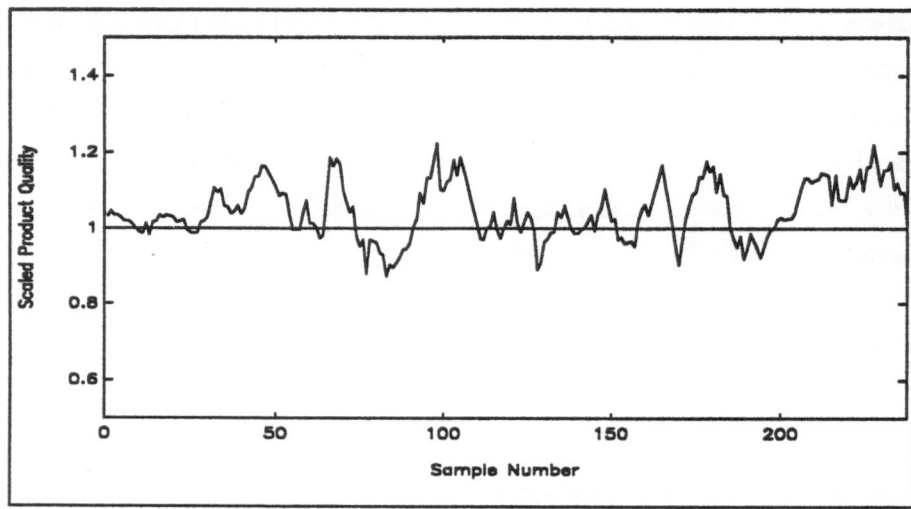

Figure 7.7   Product Quality Time Series for June 1991

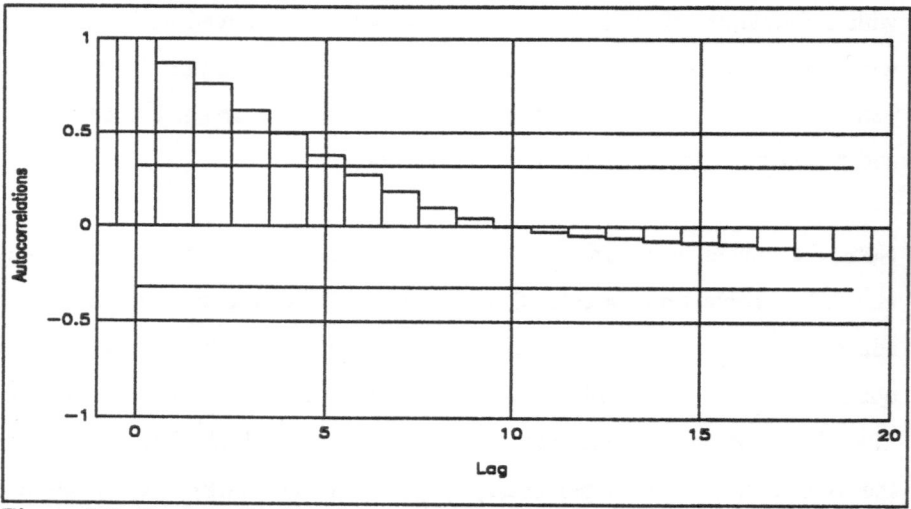

Figure 7.8   Product Quality Autocorrelation for June 1991

further steadied the control performance and eliminated any low frequency cycling.

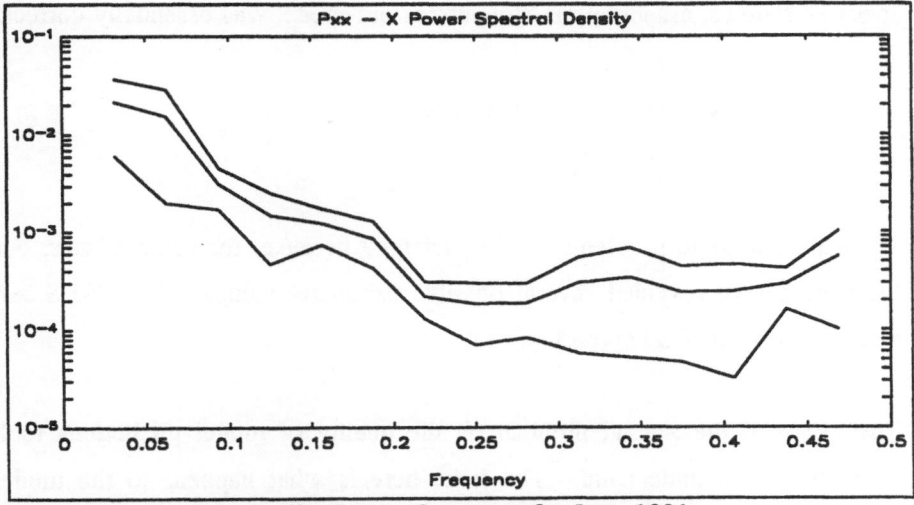

Figure 7.9 Product Quality Power Spectrum for June 1991

7.3.6 January 1992 - Model Validation

Even with this retuning, our desired level of performance had still not been achieved. In July 1991, we felt it necessary to do some closed-loop plant tests in order to verify the accuracy of our process models. Several key disturbance variables were identified and pseudo-random binary signal (PRBS) tests of these variables were run in the plant under closed-loop conditions. We then used time series analysis to check to see if there were any unmodelled effects in our model predictions.

We found several effects that were not fully accounted for by finding cross correlations between the key disturbance variables perturbed and the model residuals. After careful analysis, we summarized the reasons for prediction errors as follows, assuming the structure of our model was essentially correct:

1)   poor model parameter values
2)   measurement biases

We expected some problems of the first type because, for some effects, our literature search revealed several possible parameter values. The PRBS test results confirmed the appropriate value.

The effect of the second problem on the quality of model predictions took much longer to understand. The issue here is what happens to the model predictions when the value of the variables in the model are set from instrumentation systems which yield biased signals. An example is useful here.

Assume there exists a chemical reaction system in which the reaction rate may be increased by the addition of a reaction promotion agent, P. Furthermore, assume that the following model describes the theoretical behavior of the promoter:

$$r_{overall} = r_{base} + k[P]^2 \qquad (7.5)$$

where $r_{overall}$ is the overall reaction rate, r_{base} is the reaction rate with no promoter present, and [P] is the measured concentration of the promoter in the reaction phase. Finally, assume the values of r_{base} and k are known from laboratory experiments. Can this equation describing the promoter behavior be added directly into a process model?

166

It all depends on the accuracy of the measurement of the promoter concentration. If the measurement is exact, then as expected, there will not be problems. But what if the measurement has a bias that makes the actual concentration higher than what is measured? Clearly, the process model will consistently underpredict the reaction rate, forcing any adaptation routine to account for the process/model mismatch whenever the promoter concentration is varied. This can result in poor controller performance if the promoter effect is large compared to the base reaction rate, or if the promoter concentration is raised to a high value where the nonlinearity begins to grow.

We suspected after awhile that these effects were present in our model. The first attempt we made to overcome this problem was to refit the value of the adjustable parameter in the model, here in the example, k. Clearly, given measured values of the overall reaction rate and the promoter concentration, we can recompute the value of k to ensure the correct value of the overall reaction rate at a particular promoter concentration given the biases of the measurements. But all this does is to produce a local fit of the underlying process in which the point value is correct but the slope of the function is not. This is due to the fact that the promoter model still does not account for the concentration measurement bias.

Clearly, it is necessary to add adjustable parameters into the model in order to account for these bias problems. This goes against the philosophy often seen in linear model development of minimizing the number of adjustable parameters. However, where linear models require only the slope of a function to be specified, nonlinear models require both the slope and value to be correct. The reaction rate model in equation 7.5 could be modified to the following form:

$$r_{overall} = r_{base} + k * ([P] - [P]_{bias})^2 \qquad (7.6)$$

in order to permit better model predictions. The value of $[P]_{bias}$ can be computed from a PRBS test on the promoter concentration. Our chosen model had two major effects that required the addition of bias parameters.

The effects of correctly tuning the adaptation algorithm and accounting for measurement bias on the control performance can be seen in Figures 7.10 and 7.11, which display the controller performance for January 1992. We are now at minimum variance control, which means that the controller is moving the manipulated variable as hard as it can. At minimum variance, further improvements in controller performance will only be realized by reducing the disturbance variances.

7.3.7 March 1992 - Economic Benefits

By March 1992, we had checked and verified the majority of effects in the model, and felt a full review of the controller performance was necessary. Up to this point, the controller performance had been evaluated by analyzing the quality of the line samples. While it is indicative of performance, the plant measures its performance in terms of on-spec and off-spec product batches. Batches of product are made by segregating and blending every 4 hours of plant production. Final analysis of the product is then done on a

168

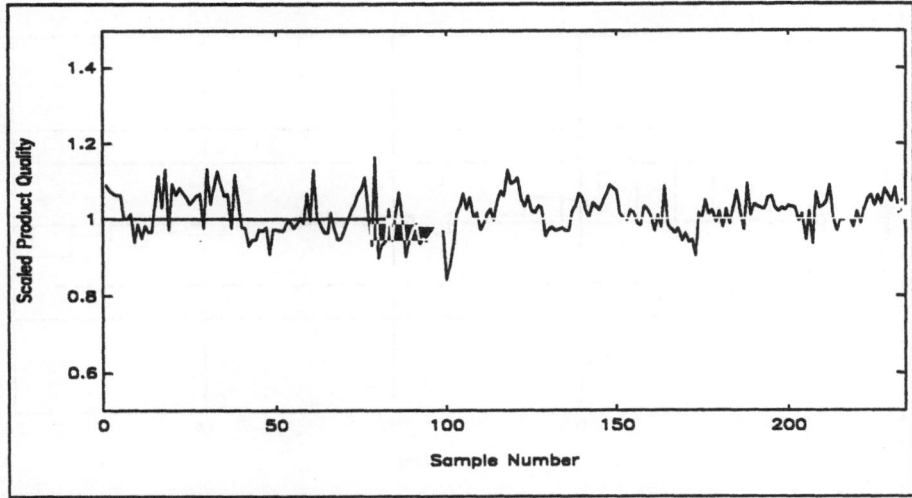

Figure 7.10   Product Quality Time Series for January 1992

composite sample of the batch.

For each month since January 1990, the percentage of on-spec batches to all batches produced in that month was computed. These percentages are plotted in Figure 7.12.

It is clear that significant improvements have been made in the plant performance since the beginning of the project.   The first improvement occurred when the controller was first put on-line in October, 1990, when the percentage on-spec increased on average from 75% to 80%.   The second improvement occurred when the adaptation mechanism was tuned properly in July 1990.   Then the performance moved up to 90% on-spec.

This improvement in on-spec percentage translates into less off-spec material

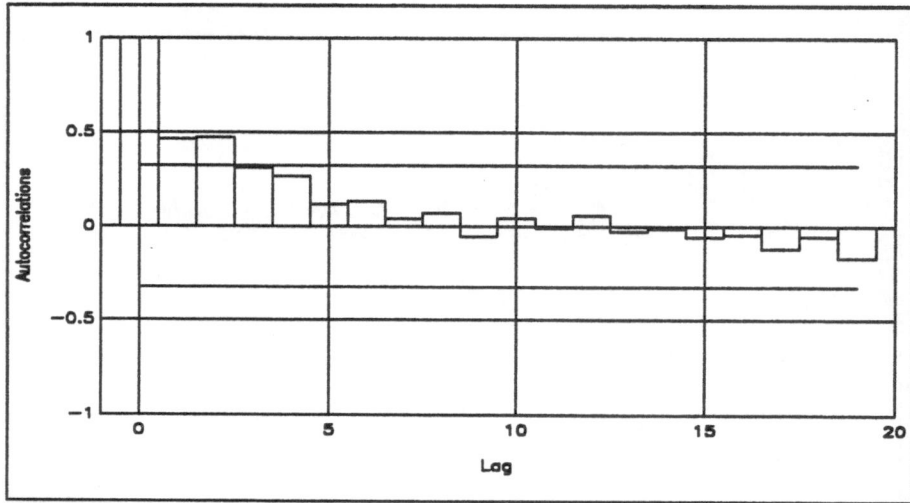

Figure 7.11   Product Quality Autocorrelation for January 1992

that must be sold at a lower price.   Another way to look at the improvement in on-spec percentage is that it decreases the production time required on the process to make a given amount of product.

Given the differential in selling price between on-spec and off-spec product and the plant throughput, this improvement in performance is worth in the region of $CDN2,000,000/year, which more than pays for the 2 man-years development time.

7.3.8 Process Monitoring

From our model validation work, we began to realize that we could recover a great deal of information about our process/model mismatch by analyzing the time series of our adaptive model parameter.   For example, it had always

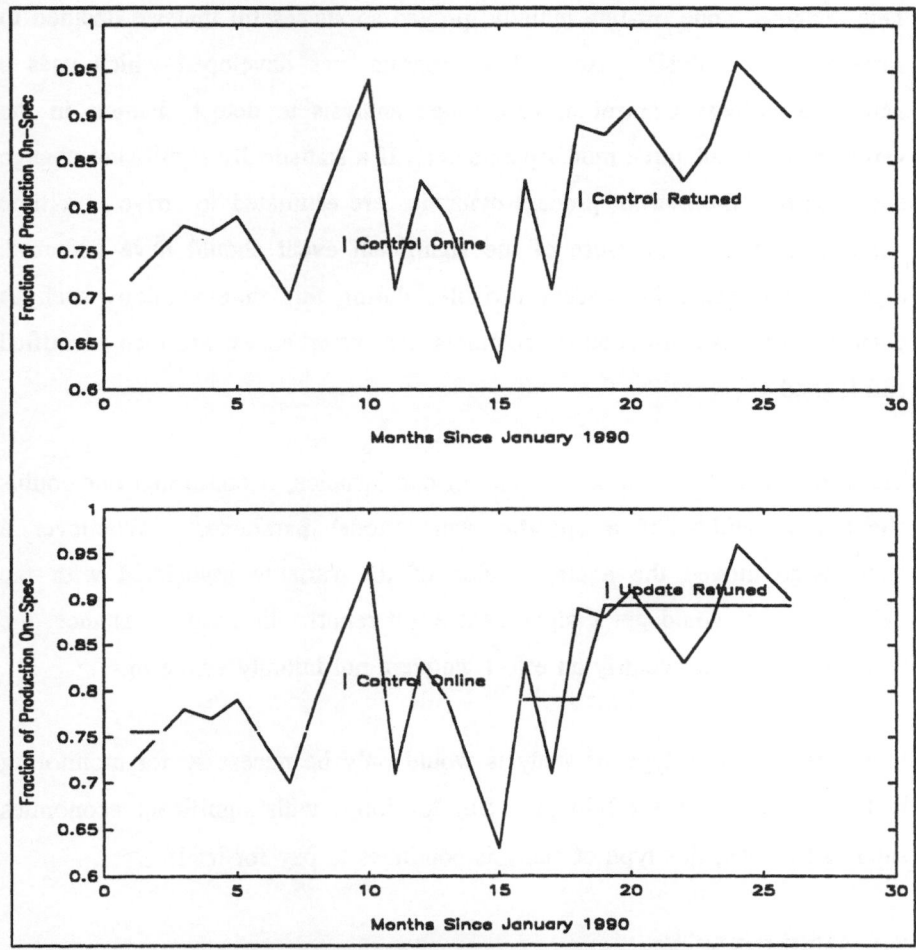

Figure 7.12 Long-Term Product Quality Control Performance

been suspected that changing levels of impurities in the reactant streams could cause the final product quality to change dramatically. We were finally able to prove this by correlating changes in the update parameter to changes in reactant impurities.

171

This analysis, done off-line initially, proved so successful that we decided to automate the analysis. An on-line program was developed which uses a Statistical Process Control moving range analysis to detect changes in the variance of the adaptive model parameter. If a statistically significant change has occurred, the current process dynamics are estimated to arrive at a time window in which the cause of the significant event should have occurred. Significant changes in process variables during this time window, such as changes in feedstream sources, setpoints and disturbances, are then identified and logged.

The program has been most useful. In one instance, it confirmed our doubts about the validity of a specific static model parameter. Whenever a disturbance moved the average value of the variable associated with the parameter, we would get a significant event report. In another instance, the report helped us to identify an effect we were not initially aware of.

It is clear that this type of analysis would only be necessary for monitoring high performance control loops. But for loops with significant economics such as this one, this type of analysis continues to pay for itself.

7.4 SUMMARY

It can be seen that the journey from original concept to actual achievement of the desired performance level has been a long and often difficult one. If one reason had to be singled out for the long development time, it would simply be the lack of experience in implementing such strategies in an industrial environment and on such a complex process. There are few published results on such issues as nonlinear model selection, propagation of measurement errors and biases through nonlinear models, and tuning rules for adaptive controllers of this form.

Having said that, we at Shell Canada believe the results were worth the effort. Certainly, the economics benefits of the control strategy have paid back the time and money expended in its development. Knowing that we are now at minimum variance control has changed our philosophy when investigating poor process performance. Before we would assume that our control moves were not appropriate for the disturbance. Now, we know that an increase in quality variable variance is due to an increase in disturbance variable variances. As a result, we have become very sophisticated in our knowledge of the disturbance variables affecting the product quality variable. We now know what our main unmeasured disturbances are, how often and how large these disturbances are, and how they affect the product quality variable. We are now in a position to prioritize these effects for incorporation into our on-line model.

7.5 ISSUES IN NONLINEAR ADAPTIVE CONTROL

There are many issues around the development of such systems that still need to be addressed. In summary, I would like to present the following comments:

1)   The key issue in any nonlinear controller development is the choice of process model, and not necessary the choice of

controller algorithm.   Given that the model structure is appropriate for GMC implementation, I have no concerns about the GMC algorithm itself. It has proven to be stable, robust and easily implemented.

2)   Process model selection is a key issue.   One must strike a balance between detail and accuracy in the model and the quality of instrumentation that will provide the inputs into the model. In order to capture important effects in the process, the instrumentation measuring the effects must be good.

3)   We have had great success in collecting any process / model mismatch in an adjustable model parameter rather than letting the error accumulate in a controller integral term. Adapting a model parameter permits constant monitoring of the model performance, and can suggest areas in which the model may be improved.

4)   On-line implementation of nonlinear models will often require the addition of extra static parameters in order to properly account for measurement biases and errors.

5)   Assuming adaptive methods are used, the GMC algorithm can be modified by eliminating the integral term.   This simplifies the tuning of the controller; the control engineer need only now specify the closed loop time constant of the system.

6)   While the GMC tuning is often straightforward, tuning of the adaptive parameter update can be difficult. Although the theory is well developed, we lack documented real-world experience in tuning these systems. We found that the use of power spectrum techniques greatly assisted in the tuning of the exponential data

174

weighting algorithm.

7)   For key effects in the model, we have had great success with performing closed-loop PRBS experiments on the process to verify model parameter values.   Often, the data quality of happenstance process upsets is not sufficient for this task.

8)   For any high performance control loop, the controller performance should be constantly measured in order to permit comparisons before and after changes to the control system. New monitoring techniques such as the autocorrelation method of Harris (1989) provide mathematically strong indicators that can be used along with company-specific performance measures.

7.6 REFERENCES

Dumont   G.A. (1982)   Self-tuning Control of a Chip Refiner Motor Load. Automatica Vol 18, 13:307.

Goodwin   G.C. and Sin K.S. (1984)   Adaptive Filtering, Prediction and Control. Prentice-Hall Inc., Englewood Cliffs, New Jersey.

Harris   T.J. (1989)   Assessment of Control Loop Performance. Can J Chem Eng 67:856

Kozub   D.J. and MacGregor J.F. (1990)   Feedback Control of Polymer Quality in Semi-Batch Copolymerization Reactors. McMaster Advanced Control Consortium Report, McMaster University, Hamilton, Canada

CHAPTER 8

BLAST FURNACE STOVE CONTROL

8.1 INTRODUCTION

Control in the minerals and metals processing industries presents many challenges to modern control algorithms. These processes have often been ignored by control engineers, and yet the incentives to improve the operation of these plants are quite large. This case study examines one such application.

8.2 BLAST FURNACE STOVES

8.2.1 Physical Description of Full-Scale Stoves

In steel production, iron ore is first reduced to iron in a blast furnace. The solid ingredients of iron ore, coke and limestone, are poured into the top of the furnace and enter the tuyères region, into which large volumes of high temperature, high pressure blast air are blown. The coke burns in this hot air blast, generating temperatures exceeding 2000°C and producing carbon monoxide gas. At such high temperatures the coke and gas react with the

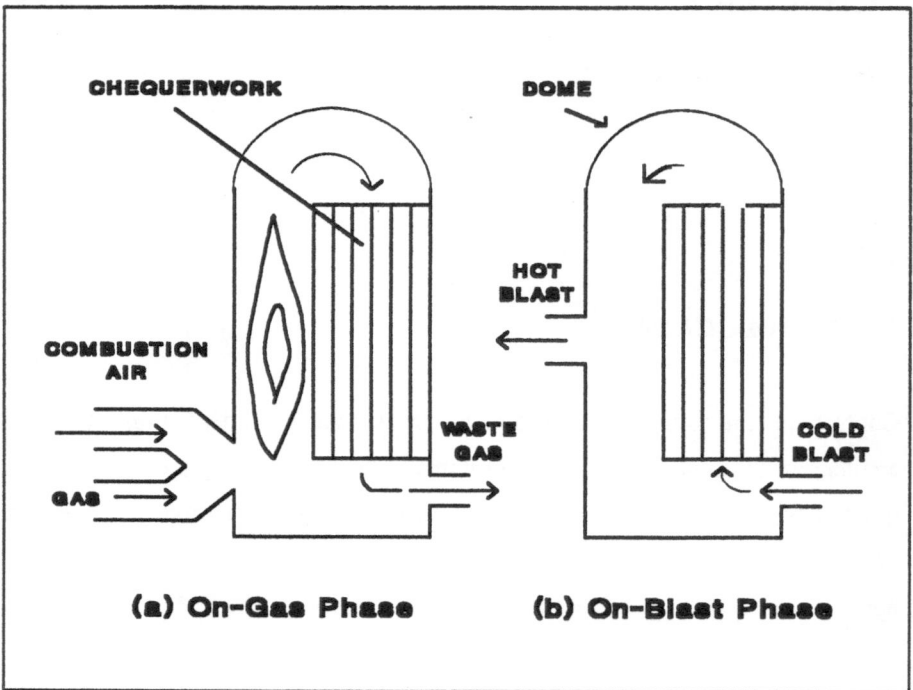

Figure 8.1 Blast furnace stove operation; a, on-gas phase; b, on-blast phase

iron oxide in the ore, reducing it to molten iron.

To ensure efficient furnace operation, the required flow-rate and temperature of the blast air must be maintained. Typical industrial flows are 3000-10000 Normal $m^3 min^{-1}$ at temperatures of 1000-1300°C, depending on the size and type of furnace (Beets et al 1977). This hot air provides up to 40% of the blast furnace sensible heat requirement (Nose et al 1984). It in turn utilizes approximately 30% of the total fuel consumption in the steel making process (Mitter et al 1981). From both operational and economic points of view

178

therefore, a reliable and controllable blast air delivery system is required.

The hot blast requirement is met through use of regenerative heat exchangers known as blast furnace stoves. These units consist essentially of a

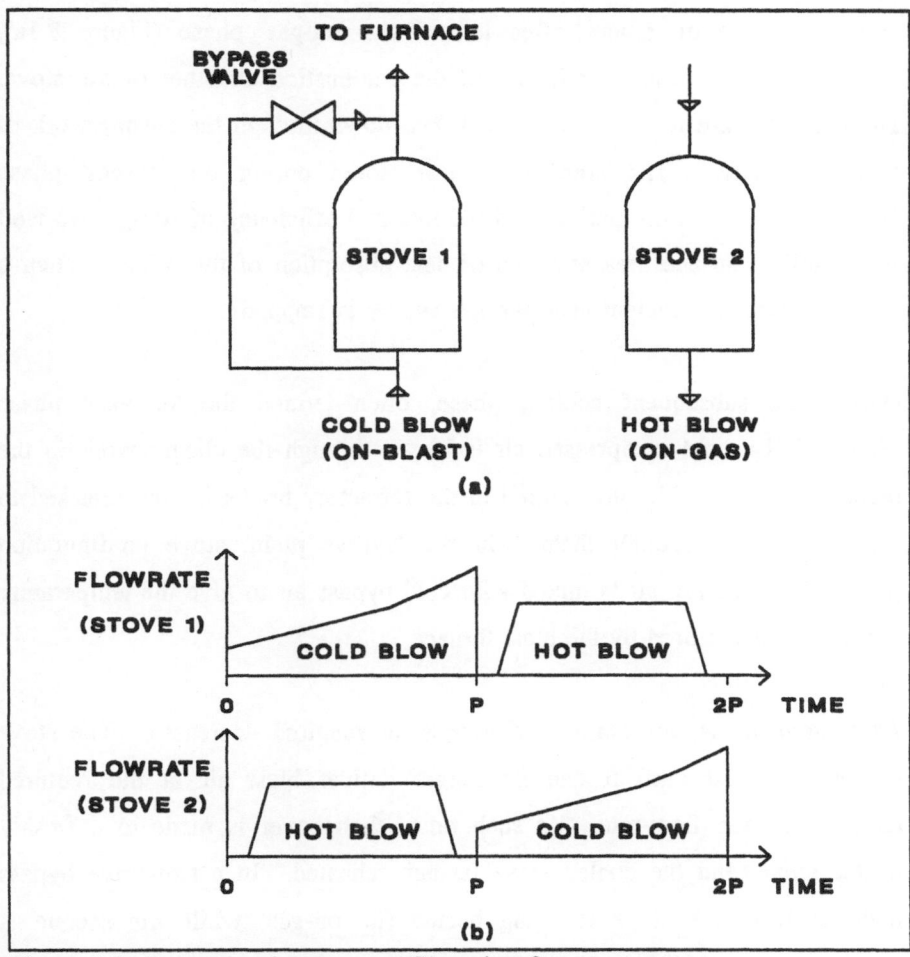

Figure 8.2 Bypass main stove configuration for two stoves

179

chequerwork of refractory brick, which performs the heat exchange function of the stove. As a regenerative heat exchanger, a stove's operation consists of two distinct phases: a heating phase to store heat in the stove, followed by a cooling phase, in which the stored heat is released to another fluid.

During the heating phase, often termed the 'on-gas' phase (Figure 8.1a), combustion gas is burned with air in the combustion chamber of the stove. The hot gases are forced upwards and then down through the chequerwork of refractory brick. The amount of heat stored during the 'on-gas' phase depends on the heating period, and the rate and efficiency of firing, was well as the mass, surface area and rate of heat absorption of the brick. When a stove has stored sufficient heat, the gas supply is stopped.

During the subsequent cooling phase, often termed the 'on-blast' phase (Figure 8.1b), cold compressed air is blown through the chequerwork in the reverse direction. The heat stored in the refractory bricks is then released to the air passing through them. In the 'bypass main' stove configuration (Figure 8.2) this hot air is mixed with cold bypass air to give the temperature and flow-rate required by the blast furnace.

As a stove cools, the amount of bypass air required decreases. The stove remains on-blast until it can no longer deliver blast air at the required temperature and flow-rate. At such time, changeover is made to a freshly heated stove, and the cooled stove is then reheated. In a two-stove bypass main system, one stove is being heated (ie. on-gas) while the second is providing blast air (ie. on-blast) (Figure 8.2).

8.2.2 Laboratory Scale Stove System

The laboratory-scale blast furnace stove system used in this investigation is shown schematically in Figure 8.3. It utilized two stoves in the bypass main mode of operation.

The brass shells of the two stoves provided the thermal capacitance for the system. A network of copper tubing and two-way solenoid valves supplied hot on-gas air to the on-gas stove, and cold on-blast air to the on-blast stove. The solenoid valves worked in pairs and were switched to allow changeover of stove duties between on-gas and on-blast roles.

The stove exit temperatures were measured by three semiconductor sensor probes. These probes were located at the stack outlet (T_1), the junction of the on-blast stove exits (T_2), and the blast outlet, downstream of the bypass air mixing point (T_3).

Supply air was regulated at 280 kPa gauge and distributed through a manifold to provide compressed air for both the process on-gas/on-blast air requirements and for pneumatic valve actuation. Air delivery to the on-blast stove was regulated by a control valve, termed the 'main valve'. A second control valve, called the 'bypass valve', controlled the bypass air flow. Together the two valves regulated the overall volumetric flow-rate and temperature of the blast air mixture. Total blast air and on-blast stove air flow-rates were measured by two differential pressure cells. Maximum on-blast and on-gas air flow-rates of 4000-5000 l h^{-1} could be supplied.

The hot air for the on-gas heating phase was delivered by an electric air heater, rather than a combustive system as in the industrial case. (In this investigation the control of the combustion process was not considered.) The

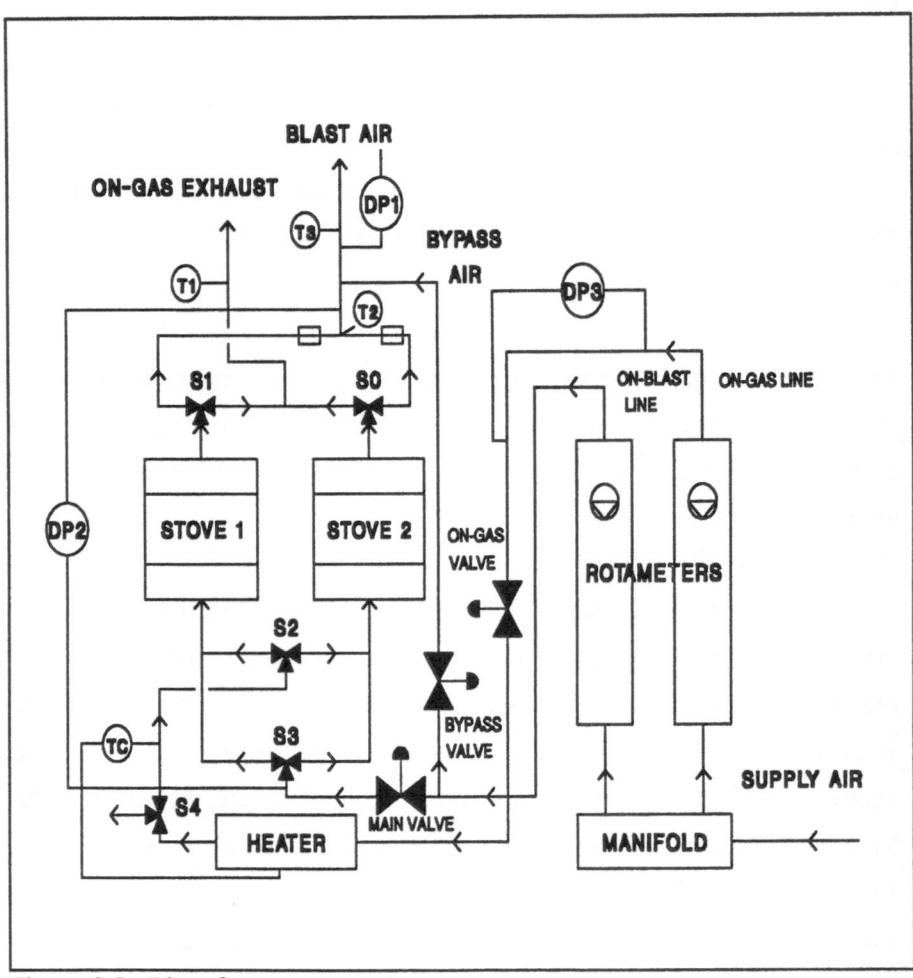

Figure 8.3 Blast furnace stove rig

heat input to the on-gas stove was varied by manipulation of the hot air temperature. A local feedback control loop on the heater allowed control of the heated air within the range of 0-100°C. Heat input to the on-gas stove can also be varied through manipulating the flowrate. A control valve was installed in the on-gas air line to allow regulation of the air flowrate. In all subsequent studies it was decided to only use the on-gas air flowrate as the manipulated input, and hold the on-gas air temperature constant. This is consistent with the aim of maximizing the thermal efficiency of the stove operation.

The temperature and flow-rate ranges of the experimental rig were well below the 1300°C and 3000-10000 m^3min^{-1} blast of an industrial stove system. In addition, the cold air in the on-blast phase flows co-currently with the hot on-gas air, rather than cooling the stove by a reverse flow of air, as in the industrial system. The experimental system was nonetheless found to reproduce the basic behaviour of the larger industrial system.

8.2.3 Control Objectives, Measurements and Manipulated Variables

The primary objective of a blast stove control strategy is to maintain the prescribed blast temperature and flow demands, while maintaining as high an overall thermal efficiency as possible. The major controlled variables are therefore the hot blast temperature and flow-rate. The associated manipulated variables are the main and bypass flow-rates in the on-blast phase, the inlet gas flow-rate and temperature in the on-gas phase, and the period of the operating cycle.

The stove control task may be divided into two general areas: short and long term control goals. Manipulation of the bypass and on-blast stove flow-rates determine the mixed temperature and flow-rate of the resulting blast air. This represents a short term control goal. Short term control is made difficult by significant process interactions, non-linearities, and time-varying behaviour. This problem will be quantified in section 8.2.4.

In the long term, proper operation requires thermally efficient control of stove heat supply and storage, in order to meet the "hot blast" demand. This is particularly true when heat input increases are required over several periods to meet a load increase. However, thermal efficiency requirements conflict with the need to maintain both cyclic stability and to obtain rapid responses to load changes.

The level of on-gas heat applied to meet a load is instrumental in maintaining *cyclic stability*. Insufficient heating for a particular blast load can bring about cyclic collapse. Collapse results when the system cannot provide, independent of the number of cycles, sufficient heat to meet a particular load. Stove changeovers would become more and more frequent. If allowed to continue unrestricted, on-blast durations in the long term would become shorter and tend to zero, as the stove system attempts to deliver a demand it is incapable of sustaining. The greater the heat input, the lower the risk of cyclic collapse.

The choices made in on-gas heating are, however, also instrumental in determining the *thermal efficiency* for the stove system. Blast stoves utilise

approximately 30% of the total fuel consumption in the steel making process
(Mitter et al 1981). Given the current high cost of energy, stove operation
has enormous consequences on the efficiency, and therefore profitability, of
steel production. Clearly, the minimisation of heat input in the stove
operation is sought for efficient operation.

At the same time, blast furnaces require *rapid responses to load changes* in
the blast air demanded. Rapid response, however, comes at the expense of
either cyclic stability or thermal efficiency, depending on how the level of
on-gas heating is manipulated. If rapid response is attempted with
insufficient heat input, cyclic collapse is risked. With high heat input,
thermal efficiency is compromised.

The conventional industrial method of stove control is that of maintaining a
fixed period of operation. Although simple and flexible, this method has
been shown to incur considerable thermal efficiency penalties (Jeffreson
1979) and slow responses to load changes. Notably, this method relies on a
high level of residual heat in the on-blast stove at the end of a period.
To summarize the above discussion, the following statements can be made:

A.   **Short-term Control:**

Control Objectives:	Total blast air flowrate (F_T)
	Blast air temperature (T_3)
Manipulated Inputs:	Main air flowrate over stove (F_g)
	Bypass air flowrate (F_b)
Measurements:	Total blast air flowrate
	Blast air temperature

185

B.   **Long-term Control:**

Control Objectives: Cyclic Stability

Thermal Efficiency

Speed of response

Manipulated Inputs: On-gas air flowrate

Blast air temperature setpoint (T_3^*)

Measurements: Total blast air flowrate

Blast air temperature

On-gas exit air temperature (T_g)

8.2.4 Control Difficulties

The short-term control task is made difficult by several factors. First, the highly coupled blast temperature and flow-rate responses exhibit significant process interactions in any traditional single-input, single-output control scheme. The relative gain array for the experimental system is:

$$\Lambda = \begin{matrix} F_b & F_g \\ \begin{bmatrix} 0.61 & 0.39 \\ 0.39 & 0.61 \end{bmatrix} & \begin{matrix} T_3 \\ F_T \end{matrix} \end{matrix}$$

where the manipulated variables are the bypass flow, F_b, and the main stove flow, F_g. The potential advantage of a multivariable control scheme is therefore quite evident. In addition, the blast temperature response demonstrates significant non-linearity and time-varying behaviour over the course of an on-blast period. This is demonstrated in Table 8.1 for the experimental system. The pseudo-steady state open loop gains between blast

186

temperature and on-blast stove flow-rate are given for step increases and decreases, both early and late in the on-blast period.

Stove changeover represents a particularly disruptive disturbance in blast stove operation. Upon changeover from a cooler stove (with zero or low bypass flow) to a hot stove (with high bypass flow), the high level of interaction between the blast flow and temperature loops becomes more evident. In addition, there is a sudden change in the process dynamics. A potentially destabilizing period of oscillation in the controlled variables may ensue.

Table 8.1 Gains for blast temperature - stove flow response

Step in on-blast stove flow	Gain ($°C(l\ h^{-1})^{-1}$) x 10^3	
	Start of on-blast	End of on-blast
Increase	2.01	0.10
Decrease	0.73	0.48

The long term control problem is made difficult by the need to balance the competing objectives of maximizing the thermal efficiency and the cyclic stability, while providing rapid response to changes in the blast load-flowrate or temperature. This type of problem is ideally suited to the application of constrained Generic Model Control as discussed in chapter 2.

8.3 MODEL AND CONTROL LAW DEVELOPMENT

8.3.1 Blast Air Temperature

A model of the blast air temperature can be developed from a simple mass and energy balance around the mixing point of the main and bypass air flowrates. Thus

$$\frac{dT_3}{dt} = \frac{1}{\tau_{T_3}} \left\{ \left(\frac{F_g}{F_T}\right) T_2 + \left(\frac{F_T - F_g}{F_T}\right) T_{IN} - T_3 \right\} \tag{8.1}$$

where T_3 is the blast air temperature (°C)

T_2 is the air temperature of air leaving the stove (°C)

T_{IN} ambient air temperature (°C)

F_g is the on-blast stove flowrate (l h^{-1})

F_T is the total blast air flowrate (l h^{-1})

and τ_{T_3} is the time constant of the blast air temperature sensor (h)

Verification of this model was performed by implementing a step change in the stove flowrate (F_g) and recording the blast air temperature T_3. This is shown in Figure 8.4. It can be seen that the model fits the recorded data quite well.

A GMC control law can be developed from equation 8.1 and a reference system performance trajectory:

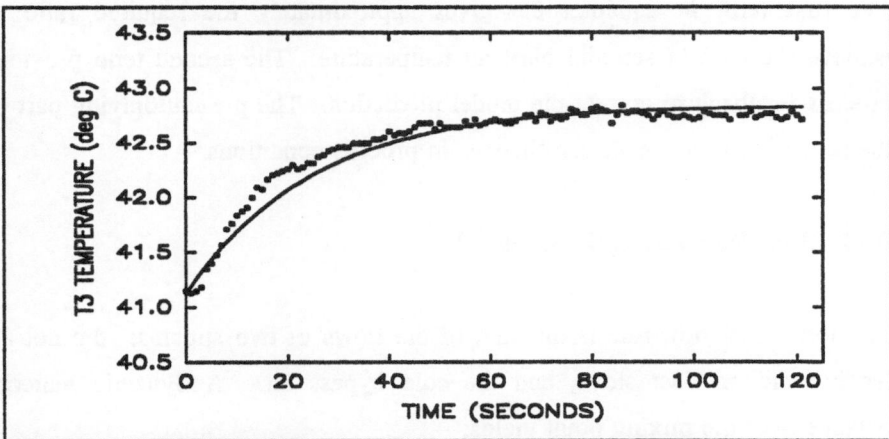

Figure 8.4 Blast temperature response to an increase in stove flow

$$\left(\frac{dT_3}{dt}\right)^* = K_{1_\tau}\left(T_3^* - T_3\right) + K_{2_\tau}\int_0^t \left(T_3^* - T_3\right)dt \tag{8.2}$$

If the ratio of the flowrates $\dfrac{F_g}{F_T}$ is used as the manipulated variable, then the

control law is:

$$\frac{F_g}{F_T} = \left(\frac{T_3 - T_{IN}}{T_2 - T_{IN}}\right) + \frac{\tau_3}{(T_2 - T_{IN})}\left\{K_{1_\tau}\left(T_3^* - T_3\right) + K_{2_\tau}\int_0^t \left(T_3^* - T_3\right)dt\right\} \tag{8.3}$$

This ratio $\dfrac{F_g}{F_T}$ is used as a setpoint R^* to a low level PID controller to be

discussed in section 8.4. The low level PID controller would achieve this
ratio R^* by adjusting the bypass flowrate.

The first term in equation 8.3 gives approximately the required ratio to achieve the desired setpoint blast air temperature. The second term provides process feedback to correct the model prediction. The pre-multiplying part of the second term accounts for changes in process conditions.

8.3.2 Blast Flow-rate GMC Control Law

The total blast flow-rate is the sum of the flows of two streams: the hot air leaving the on-blast stove, and the cold bypass air. A dynamic material balance over the mixing point yields:

$$\frac{dF_T}{dt} = \frac{dF_g}{dt} + \frac{dF_b}{dt} \tag{8.4}$$

Equation 8.4 is in a convenient form for substitution into the general GMC control law.

$$\begin{aligned} \frac{dF_T}{dt} &= \frac{dF_g}{dt} + \frac{dF_b}{dt} \\ &= \{K_{1,r}(F_T^* - F_T) + K_{2,r}\int_0^t (F_T^* - F_T)dt\} \end{aligned} \tag{8.5}$$

If the total blast flow-rate is controlled by manipulating the on-blast stove flow-rate, F_g, then rearrangement of equation 8.5 to isolate the stove flow term, dF_g/dt, gives equation 8.6.

$$\frac{dF_g}{dt} = \frac{-dF_b}{dt} + \{K_{1,r}(F_T^* - F_T) + K_{2,r}\int_0^t (F_T^* - F_T)dt\} \tag{8.6}$$

Finite differences were used to approximate the continuous derivatives in equation 8.6. In the discrete terms of a digital sampling system, the equation becomes:

$$\frac{(F_{g_{(k+1)}} - F_{g_k})}{T} = \frac{-(F_{b_{(k+1)}} - F_{b_k})}{T} + PI_F \tag{8.7}$$

where

$$PI_F = K_{1_F}(F_T^* - F_{T_k}) + K_{2_F} T \sum_{i=0}^{k} (F_T^* - F_{T_i}) \tag{8.8}$$

The bypass air flow-rate F_b, was ot measured directly on the rig system. A suitable substitution was therefore determined. Since the temperature control law, equation 8.3 utilized the ratio of stove flow to blast flow, it was decided to introduce this ratio into the blast flow control law as well. Substituting the ratio of stove flow to blast flow, $R = F_g/F_T$, into a steady-state material balance and rearranging gives the bypass flow, F_b:

$$F_b = F_g(1/R - 1) \tag{8.9}$$

Eliminating F_b from equation 8.7 then yields equation 8.10.

$$F_{g_{(k+1)}} = R_{(k+1)}\{F_{T_k} + TPI_F\} \tag{8.10}$$

Equation 8.10 represents a control law for the blast flow-rate. It determines the stove flow-rate, F_g, necessary to achieve the blast flow-rate set-point. This stove flow-rate would serve as a suitable set-point, F_g^*, for a low level PID controller manipulating the main valve.

191

Control actions of the blast temperature GMC controller, equation 8.3, would cause disturbances in the blast flow-rate. Feedforward compensation for these disturbances can be made in the flow control law, equation 8.10, by incorporating the flow ratio set-point, R^*, of the temperature control law. In the calculation procedure, therefore, the temperature control law would be computed first to yield R^*. Then R^* could be substituted into equation 8.10 for $R_{(k+1)}$.

The blast flow control law now appears as:

$$F_g^* = R^* \{F_{T_k} + TPI_F\} \qquad (8.11)$$

However, it is not possible that the flow ratio set-point, R^*, can be achieved by the low level PID controller within a single sampling interval. It is reasonable to assume that the actual flow ratio at the end of the next sampling interval will lie between the current flow ratio, R_k, and the new ratio setpoint, R^*. A filtering of the desired ratio, R^*, with the current flow ratio, R_k, could be performed to estimate a more practical ratio for feedforward use. This ratio, R_{filt}, is given by:

$$R_{filt} = \alpha R^* + (1 - \alpha)R_k \qquad (8.12)$$

where α is the filter factor.

The blast flow-rate GMC control law now becomes:

192

$$F_g^* = R_{filt}(F_{T_k} + TPI_F) \qquad (8.13)$$

For most of the stove's operation the controller's task will be solely that of disturbance rejection.   The set-point, R^*, would therefore not differ significantly from one sample time to the next, and would likewise be nearly equal to R_k at all times.   After a stove changeover or a load change, the set-point ratio R^* might change markedly from one sample time to the next.   In addition, it would differ, as described above, from the actual flow ratio, R_k. These differences, however, would diminish quickly as blast set-points were attained, and soon R^* and R_k would again be nearly equal.   The approximation made in equation 8.12 is therefore, a reasonable measure to account for lag in the low level controllers during instances of significant changes in R^*.

Experiments were performed with the GMC controller over the full range, 0 to 1, of the filter factor, α. A value of 0.75 was found to yield the best controller response.

The control action obtained from the PI_F term in equation 8.13 would be significantly affected by the magnitude of the flow ratio, R_{filt}. This would be particularly true near the beginning of an on-blast stage, when the flow ratio would be very low. At such a time, the control action given by the PI_F term would be considerably reduced.   As a result, the control law given by equation 8.13 would exhibit very poor robust properties.   It would rely largely on the $R_{filt}F_{T_k}$ model term for its control action. · Controller

193

performance would therefore be very sensitive to model errors, as the correction through the error feedback term $F_T^* - F_{T_k}$, in the PI_F term, would be reduced. An alternative formulation of the control law was therefore developed:

$$F_g^* = R_{filt}(F_{T_k}) + TPI_F \qquad (8.14)$$

8.3.3 On-gas Stove Model

To enable the development of the long term controller using constrained Generic Model Control, a model of the gas exit temperature, T_1, from the stove on-gas was required. This model was developed by considering an energy balance on the stove and resulted in:

$$T_1^{k+1} = \left(1 - \frac{T}{\tau_{T_1}}\right)T_1^k + \frac{T}{\tau_{T_1}}\left[\left(1 - \frac{T}{\tau_{T_1}}\right)T_g^{k-1} + \left(\frac{kT}{\tau_{T_1}}\right)T_{hsp}^{k-1}\right] \qquad (8.15)$$

where T_1^{k+1} is the predicted value of the sensor reading of the on-gas stove exit air temperature one time step in the future (°C)

   T_1^k is the current temperature sensor reading of the on-gas stove exit air temperature (°C)

   T_g^{k-1} is the actual air temperature from the exit of the on-gas stove one sample time ago (°C)

   T_{hsp}^{k-1} is the inlet air temperature to the on-gas stove one sample time ago (°C)

K, τ_{T_1} are physical constants determined by experiment

T   is the controller sampling time

The initial conditions for equation 8.15 are:

$$T_g^{k=0} = T_{g_1} \tag{8.16a}$$

and

$$T_1^{k=0} = T_{1_{ad}} \tag{8.16b}$$

where T_{g_1} is the theoretical temperature of first stove exit gas after
                        start of on-gas period.

            $T_{1_{ad}}$ is the temperature of the exit gas as recorded by the sensor
                        at the end of the previous cycle.

The value of T_g can be predicted from the energy balance on the stove as:

$$T_g^k = \left(1 - \frac{T}{\tau}\right)T_g^{k-1} + \frac{kT}{\tau}\, T_{hsp}^{k-1} \tag{8.17}$$

Figure 8.5 shows a plot of the model prediction versus experimental data. As can be seen, a good agreement is obtained between the model described by equations 8.15 to 8.17 and the experimentally recorded data.

An even more telling test of model accuracy is the ability to predict the cycle length over many cycles. This is shown in Figure 8.6 over 15 periods. Very good agreement is displayed throughout, even following the load increase in period 5, and the subsequent transient. This was essential to the correct functioning of the optimization procedure to be discussed in the following section.

195

8.3.4 Long-term Control Algorithm

As was discussed in section 8.2.3, the long-term control of the blast furnace stoves involves a trade-off between the thermal efficiency, cyclic stability and response to load changes. This obviously suggests that the problem should be formulated as an optimization problem, and in this work the framework of

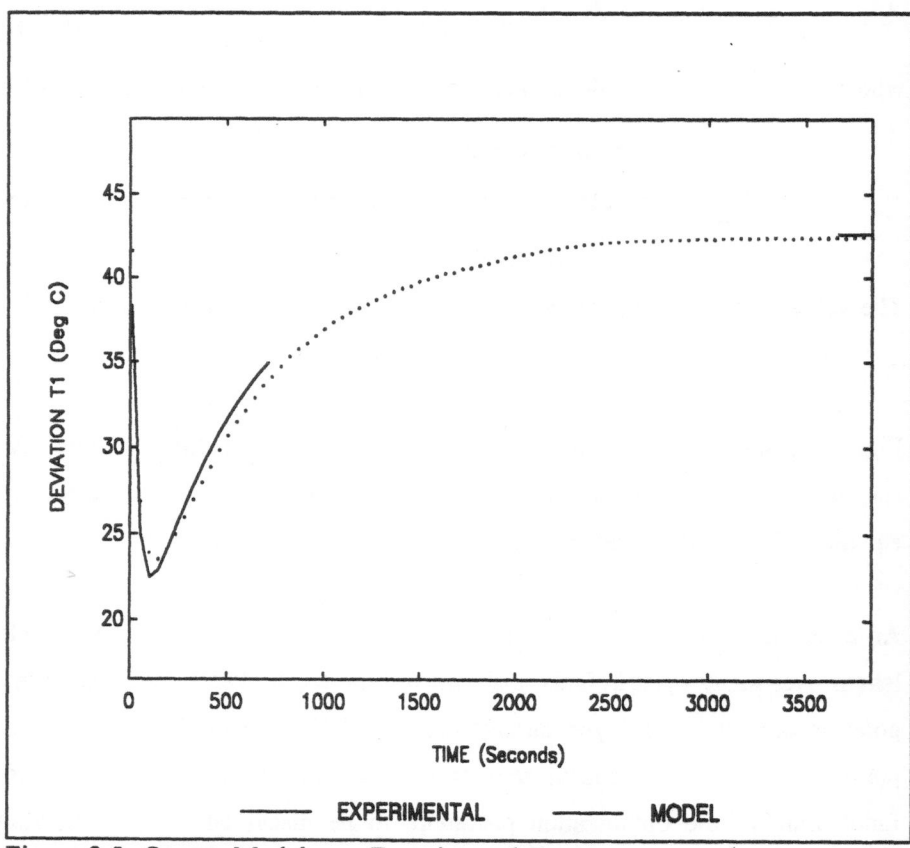

Figure 8.5 On-gas Model vs Experimental Data

196

the constrained version of Generic Model Control was used.

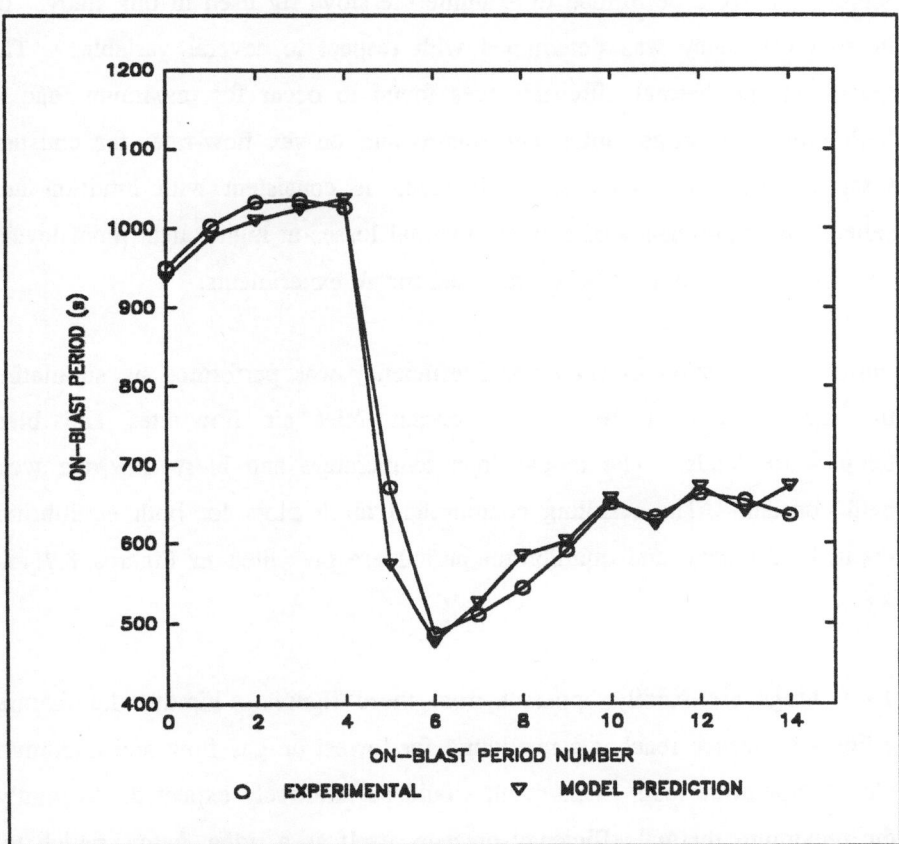

Figure 8.6 Dynamic Model - Experimental Agreement for On-blast Period Duration over a 15 Period Optimal Control Run

8.3.4.1 Limiting case operation

Simulations were performed to examine the stove rig used in this study. Its thermal efficiency was determined with respect to several variables. The maximum rig thermal efficiency was found to occur for maximum load at both minimum on-gas inlet temperature and on-gas flow-rate, for constant blast and on-gas flow-rates. This result is consistent with intuition and reflects the significance of ambient thermal losses at higher heat input levels. An inlet heat setpoint of 90°C was used for all experiments.

Further examination of rig thermal efficiency was performed by simulating the rig operation under various on-gas inlet air flow-rates and blast temperature loads. The on-gas inlet temperature and blast flow-rate were held constant. The resulting contour and mesh plots for both equilibrium thermal efficiency and equilibrium period are presented in Figures 8.7 and 8.8.

Two things are readily apparent from these figures. Firstly, the thermal efficiency surface reaches a maximum for lowest on-gas flow and maximum blast temperature load. This result would be intuitively expected. Secondly, the maximum thermal efficiency presents itself as a ridge, below which the thermal efficiency falls to zero. This corresponds to the verge of cyclic collapse. As the thermal efficiency increases in Figure 8.7, a corresponding decrease in period is shown in Figure 8.8. Where the thermal efficiency falls to zero is the point where the period falls to zero - namely, the point of cyclic collapse. The on-gas heat input has been reduced to the point of a

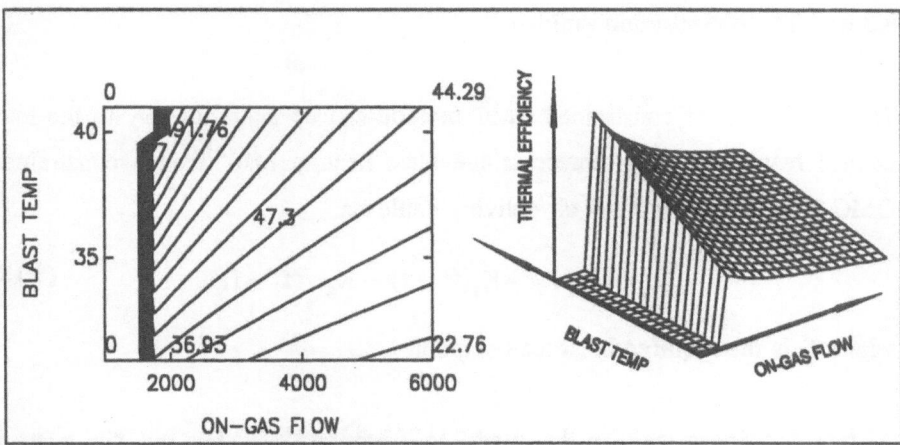

Figure 8.7   Thermal Efficiency over a range of inlet On-gas Flow-rates and Blast Temperature Loads

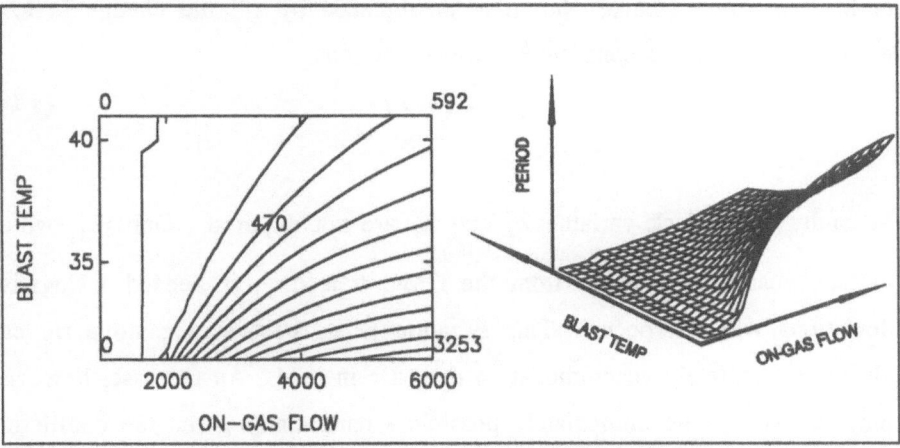

Figure 8.8   Equilibrium Period over a range of inlet On-gas Flow-rates and Blast Temperature Loads

zero-period steady state heat exchanger. This limit was described by Jeffreson (1979).

199

8.3.4.2 The optimisation problem

The application of constrained GMC method can be made directly to the load control response. If f represents the blast heating load, then a constrained GMC control law for blast air delivery could be:

$$\frac{df}{dt} + \lambda_f^+ - \lambda_f^- = K_1(f^* - f) + K_2 \int_0^t (f^* - f)dt \qquad (8.18)$$

where f^* is the required blast load setpoint

A few modifications can be made to equation 8.18 for the rig system. Firstly, the long term rig operation can be regarded in discrete elements of on-blast period.   Hence, df/dt may be replaced by Δf, the change in load delivery between two consecutive on-blast periods.

$$\Delta f^k = f^k - f^{k-1} \qquad (8.19)$$

Secondly, both slack variables λ_f^- and λ_f^+ are not required.   Only λ_f^+, which defines positive deviations from the GMC trajectory, is needed. Negative deviations will not occur. This situation exists because the stove rig can always immediately accommodate a decrease in load. An increase, however, may not always be immediately possible - particularly given the conflicting goals of thermal efficiency and cyclic stability also being considered. Therefore, only a single slack variable defining positive deviations is necessary.

The right hand side of equation 8.18 was simplified to give a control trajectory proportional to the current deviation from the desired blast load. In discretised form equation 8.18 becomes:

$$\Delta f^{k+1} + \lambda_f = K_f(f^* - f^k) \tag{8.20}$$

where f^{k+1} is the load to be delivered in the next on-blast period. K_f was specified to 1.0 so that the delivery of the full desired load is sought.

The maximum value of the slack variable λ_f^+ is the departure from the GMC trajectory which occurs when no move is made towards a new setpoint, ie. $f^{k+1} = f^k$ and $\Delta f^{k+1} = 0$. For this instance, define:

$$\lambda_{f_{max}}^+ = K_f\,(f^* - f^k) \tag{8.21}$$

The system's departure from the GMC trajectory may now be normalised by:

$$\bar{\lambda}_f = \frac{\lambda_f^+}{\lambda_{f_{max}}^+} \tag{8.22}$$

Any minimisation of $\bar{\lambda}_f$ will improve the system's response to a load change.

With a control law developed for the blast load control response, consideration will now be given to the control objectives of stability and thermal efficiency.

Firstly, given the importance of avoiding cyclic collapse for the blast rig

201

system, it would be of practical interest to know the stability margin which exists for any choice of operating variables. Equilibrium period was a good indication of system proximity to collapse. The smaller the equilibrium period, the closer the system was to the point of cyclic collapse. The absolute point of collapse was the limiting or maximum thermal efficiency situation, as described above.

Let the equilibrium period for any particular set of operating conditions be defined by P_∞. Let the equilibrium period for the same load, but at the heat input of the limiting/maximum thermal efficiency case, be P_L. A normalised period can be defined:

$$\bar{P} = \frac{P_L}{P_\infty} \tag{8.23}$$

Any attempt to minimise $\bar{P}$, ie. maximise P_∞, will increase the stability of the blast stove rig system.

Analogously, the rig's thermal efficiency may also be related to the limiting case/maximum thermal efficiency case. Let η be the thermal efficiency associated with a particular set of operating conditions. Let η_L be the limiting case thermal efficiency - the maximum thermal efficiency obtainable for the same load. Then, the normalised thermal efficiency is:

$$\bar{\eta} = \frac{\eta_L}{\eta} \tag{8.24}$$

Therefore, any minimisation of $\bar{\eta}$ will maximise the thermal efficiency.

A formal description of the blast stove rig optimisation problem may now be presented.

Blast Stove Rig Optimisation Problem

Given: Desired Blast Air Load

Choose: Deliverable Blast Air Load

                          On-gas Heat Input

To Minimise:

$$J = W_f \cdot (\bar{\lambda}_f)^2 + W_s \cdot (\bar{P})^2 + W_t \cdot (\bar{\eta})^2 \qquad (8.25)$$

where:

$$\Delta f^{k+1} + \lambda_f = K_f(f^* - f^k) \qquad (8.26)$$

$$\lambda_{f_{\max}} = K_f (f^* - f^k) \qquad (8.27)$$

$$\bar{\lambda}_f = \frac{\lambda_f}{\lambda_{f_{\max}}} \qquad \bar{P} = \frac{P_L}{P_\infty} \qquad \bar{\eta} = \frac{\eta_L}{\eta} \qquad (8.28)$$

$$u_L < u < u_U \qquad (8.29)$$

Selection of the weighting parameters, W, will allow trade-offs between response speed, stability, and thermal efficiency considerations. Quadratic functions are well known to be better suited for optimisation studies, and as such the terms of the objective function have each been squared.

8.4 IMPLEMENTATION

Implementation of both the short and long term controllers involved both a hierarchy of functions and hardware and software constructions.

8.4.1 Functional Hierarchy

The functional hierarchy is shown in Figure 8.9. The lowest level of this hierarchy is the blast furnace stoves themselves. Directly above the physical equipment, including the sensors and valves, are two "low level" PID

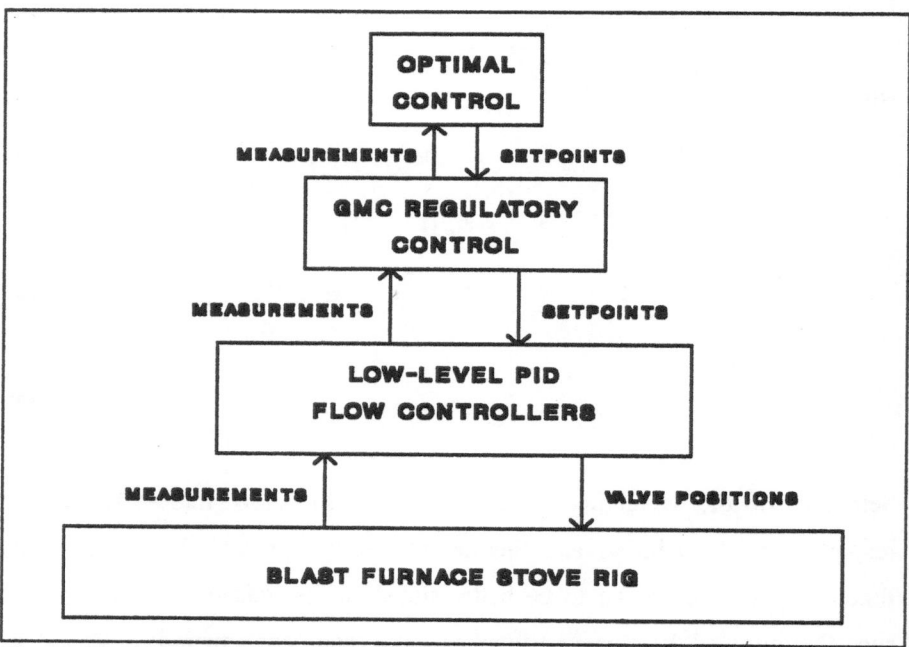

Figure 8.9 General Schematic of the Control Strategy

controllers. One of these PID loops controls the stove flowrate by adjusting the main control valve, while the other loop controls the ratio between the stove and bypass flowrates by adjusting the bypass valve.

Low level PID control loops were implemented to increase the reliability of the overall control structure. Should any of the higher level functions fail to operate, the low level PID loops would continue to maintain the last setpoint.

The setpoints for the PID loops were calculated from the GMC control laws developed in sections 8.3.1 and 8.3.2. This is shown in Figure 8.10. Finally, the long-term controller adjusts the setpoints for the GMC controllers and the setpoint of a low level controller controlling the on-gas flowrate.

8.4.2 Hardware and Software Hierarchy

The computer control hardware used to implement both the short and long term control is shown in Figure 8.11. The functions of each hardware component is shown in Figure 8.12.

The functions of each computer facility is as follows:

QNX Data Logging Server: The primary purpose of the QNX network and two PC's is to accept the sensor readings from the physical equipment and return signals to adjust the valve settings. The QNX server system is equipped with analog-to-digital as well as digital-to-analog converters, and digital input and output facilities. It communicates via a serial

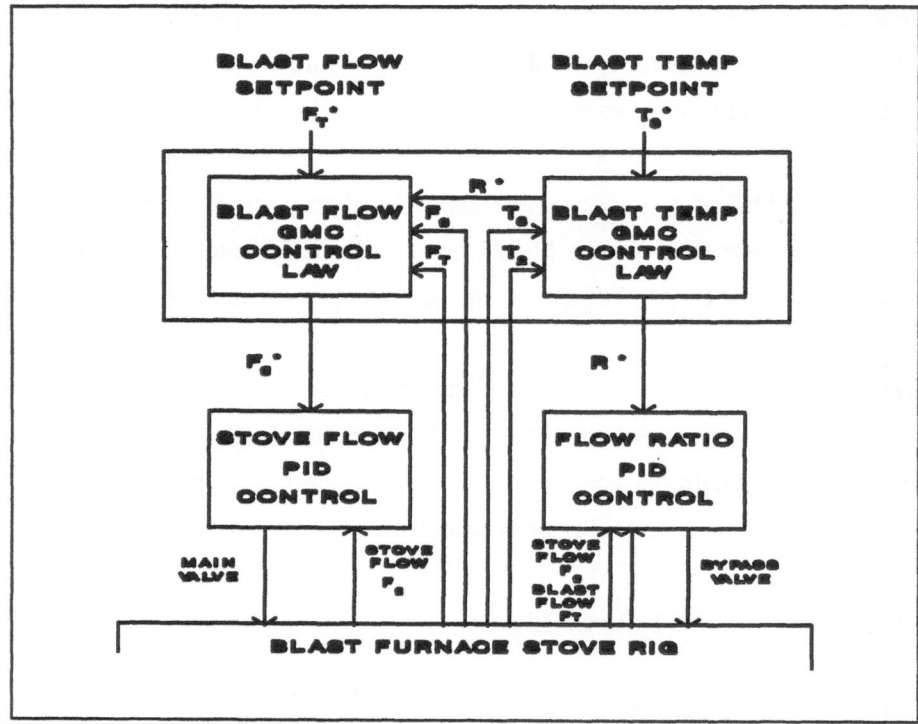

Figure 8.10 Signal flow diagram for GMC control of the blast stove rig

communication line to the TACTICIAN control system.

TACTICIAN Control System: The TACTICIAN Process Control System
    (Turnbull Control Systems, 1988) is a real-time personal computer based
    system. It allows the design and execution of process control strategies,
    simulations, and data acquisition.

The software is based on a graphic, mouse-driven user .interface that
    allows the design of executable control strategies and operator interface

206

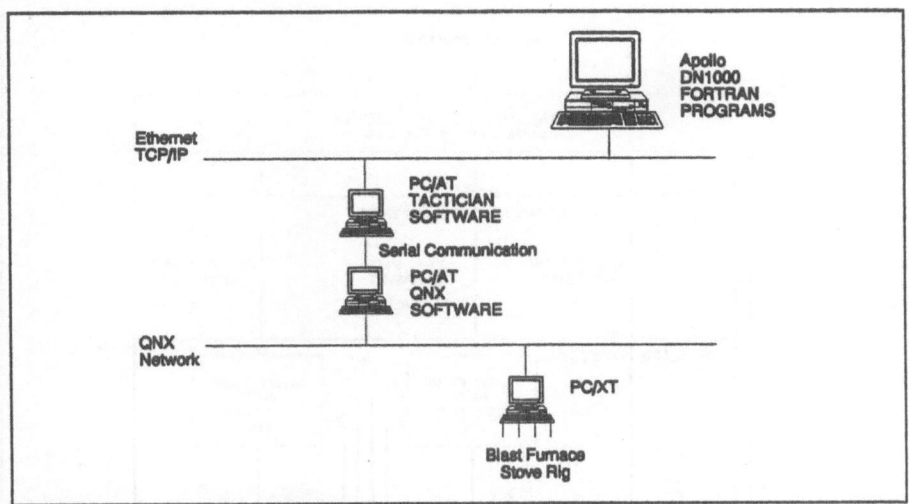

Figure 8.11 Computer Hardware Implementation

graphics.   A library of commonly used "user configurable" blocks is
available.   These encompass real-time input/output signalling, control
(eg. PID), signal conditioning, timing, basic logic, recipe loading, and
data collection and storage.   These blocks may be linked as desired to
form the required process control configuration.   For more sophisticated
functions than those provided for in the basic block library, the
developer may write C-language programs in seven "user blocks" and a
single "user task".   For instance, tasks in this investigation such as
setpoint sequencing, tailored data acquisition, complex process
simulators, and the model-based controllers required these programs.
Such programs are linked into the control configuration, and are
executed once each scan interval during TACTICIAN applications.

One particular program, the "USERCOMM" program, contains the C-

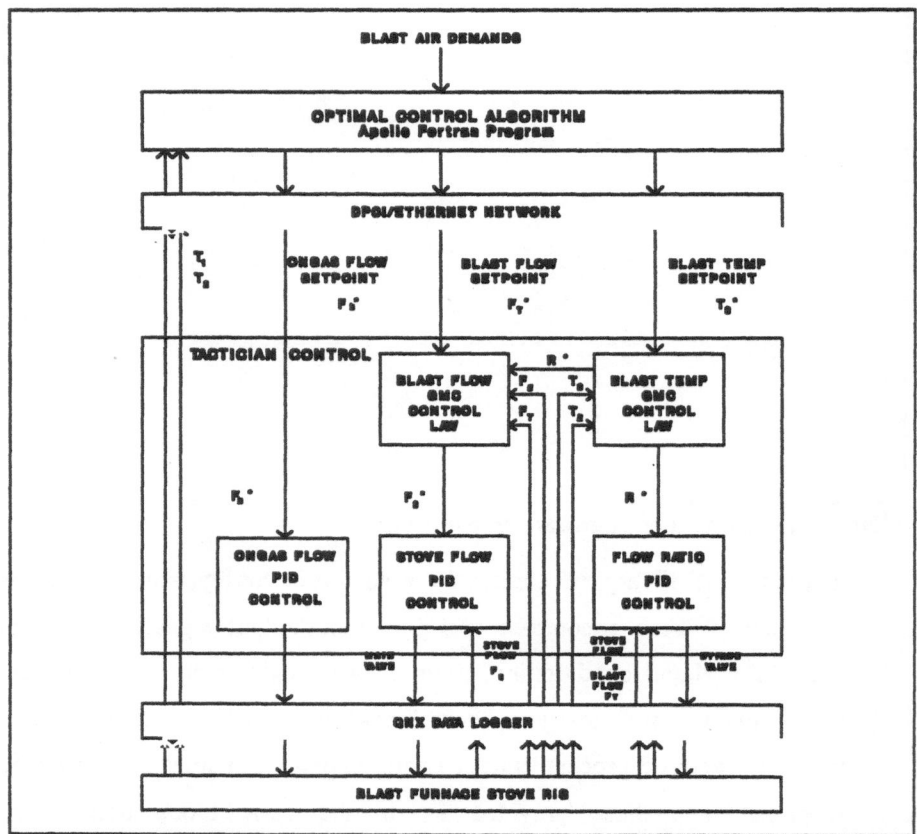

Figure 8.12 Schematic for the Optimal Control of the Blast Stove Rig

language instructions for reading and writing TACTICIAN requests and data to the serial port of the TACTICIAN computer. This accomplishes the TACTICIAN-side of communication with the QNX data logging server. It should be noted that the functions of the QNX data logging server and the TACTICIAN control system could have been implemented all within the TACTICIAN system itself. Reasons outside this single project dictated this architecture.

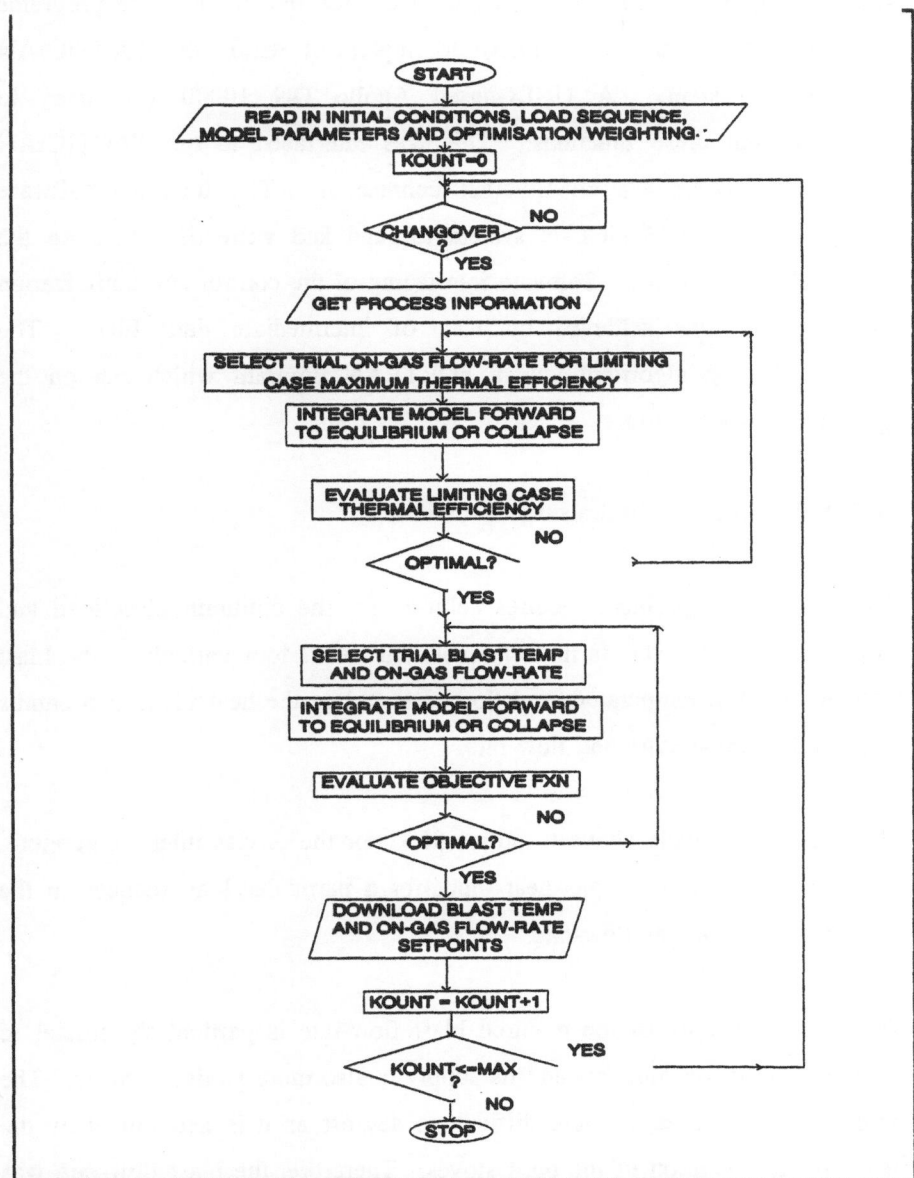

Figure 8.13 - Flowchart of Optimisation Program

Apollo Computer System: The long term control and optimization programs were too numerically intensive to implement within the TACTICIAN control system. A UNIX-based Apollo DN 10000 was used to implement these functions. This was interfaced to the TACTICIAN control system via an Ethernet connection. The interface software allowed the TACTICIAN system to read and write data files on the Apollo workstation. Software interfacing of the control and optimization functions was achieved by way of intermediate data files. The optimization algorithm was a FORTRAN program which ran on the Apollo workstation.

8.4.3 Solution of the Optimization Problem

The optimisation problem requires solution for the optimum blast load and on-gas heat input. This is in fact a solution set of four variables: the blast load entails blast temperature and flow-rate; while the heat input also entails on-gas inlet temperature and flow-rate.

The choice was made above to utilise 90°C for the on-gas inlet temperature. Hence, the choice of on-gas heat input for a particular load reduces to the selection of the on-gas flow-rate.

The constant supply of the required blast flow-rate is particularly crucial to the operation of the furnace, and its supply is also more straightforward. The temperature, however, is more difficult to address as it is determined by the heat storage condition of the blast stoves. Therefore, the blast flow-rate was always delivered, and the variable to be controlled was its temperature.

The optimisation problem therefore reduces to the selection of two independent variables: on-gas flow-rate and blast supply temperature.

Load changes were only implemented at the start of an on-blast period. Therefore, the optimisation strategy would sample the stove conditions at the end of a period and proceed to determine optimal settings for the new period. The new setpoints were then downloaded to the regulatory controllers. In addition, only load increases were considered of interest in this study. Load increases placed serious consequences on response speed, thermal efficiency, and cyclic stability. Load decreases were considered trivial as they could be met immediately and completely by the stove system at any time.

The optimisation strategy employed the dynamic models of the stove rig to simulate operation of the blast rig. The rig's operation was simulated for each set of independent variables trialled in the optimisation procedure. The initial rig conditions were specified to be those existing at the end of the preceding period, and simulation would continue forward until a cyclic equilibrium or collapse occurred. The objective function and all associated equations presented above were evaluated at the predicted eventual system conditions. Based on the successive evaluations of the objective function, the search strategy would revise its solution set and continue searching until the specified tolerance was met. Figure 8.13 presents a flowchart of the optimisation procedure.

The optimisation algorithm employed for rig optimal control was the Box Complex Method (Box 1965). This is a direct-search, non-gradient based algorithm. The Box Complex method was found to be particularly robust for this optimal control problem. This method has been successfully used for other similar industrial optimisation problems, including the optimisation of a copper anode furnace operation by Bateman (1992).

The optimisation program was executed once every period, immediately after changeover and after the process measurements from the end of the preceding period were read. Clearly, no new optimal settings could be down-loaded to the regulatory controllers until the optimisation was complete.

To ensure that the optimisation did not take an excessively long time, and consequently an overly large proportion of the actual on-blast period, a maximum was placed on the number of iterations the search could make. The maximum value used was 15 iterations, and this typically entailed a computational waiting time of 30 seconds.

This limit on the number of search iterations was generally sufficient to determine the optimum for most optimisation searches for the stove rig. In the initial instance of a load increase the search would occasionally not reach all the distance to the true optimum. However, it would closely approach it. The optimisation search of the ensuing period would then carry the rig system the rest of the way to the optimum.

Figure 8.14 shows the approximate typical scheduling for the optimal

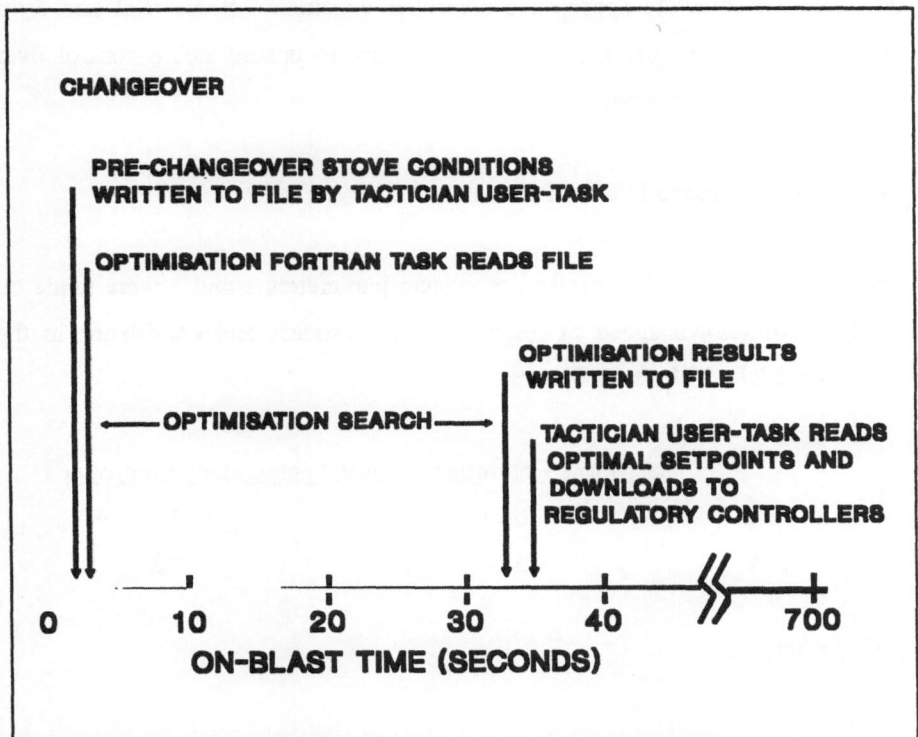

Figure 8.14 Typical Scheduling for Online Optimal Control

controller.

8.4.4 Tuning

(a) PID Loops

Open-loop step tests were performed on all PID loops, and the resultant
transient was fitted to a first-order plus deadtime model. Minimum Integral

213

Error Tuning formula for setpoint changes (Smith and Corripio, 1985) were then used to calculate appropriate controller constants. Some trial and error adjustment of these parameters were necessary to ensure stable control over the entire range of operating conditions.

(b)   GMC Reference System Parameters

The selection of the GMC reference system parameters τ and ξ were made on the basis of known speed of response of the process and confidence in the process model. Values used were:

	Blast Flow Controller	Blast Temperature Controller
ξ	10	1
τ	15	20

8.5 RESULTS

8.5.1 Short-term Results

To compare the results obtained using Generic Model Control, two PID controllers were also implemented. These controllers controlled the blast flowrate by manipulating the main valve, and the blast temperature by manipulating the bypass valve. This was the recommended pairing as suggested by the relative gain array analysis in section 8.2.4. These controllers were tuned by the techniques discussed in the previous section, and implemented on the TACTICIAN control system.

Sequences of step changes in blast temperature and flow-rate set-points were used to test the different control methods. The integral of the time-weighted absolute error (ITAE) for each step change response was used to compare

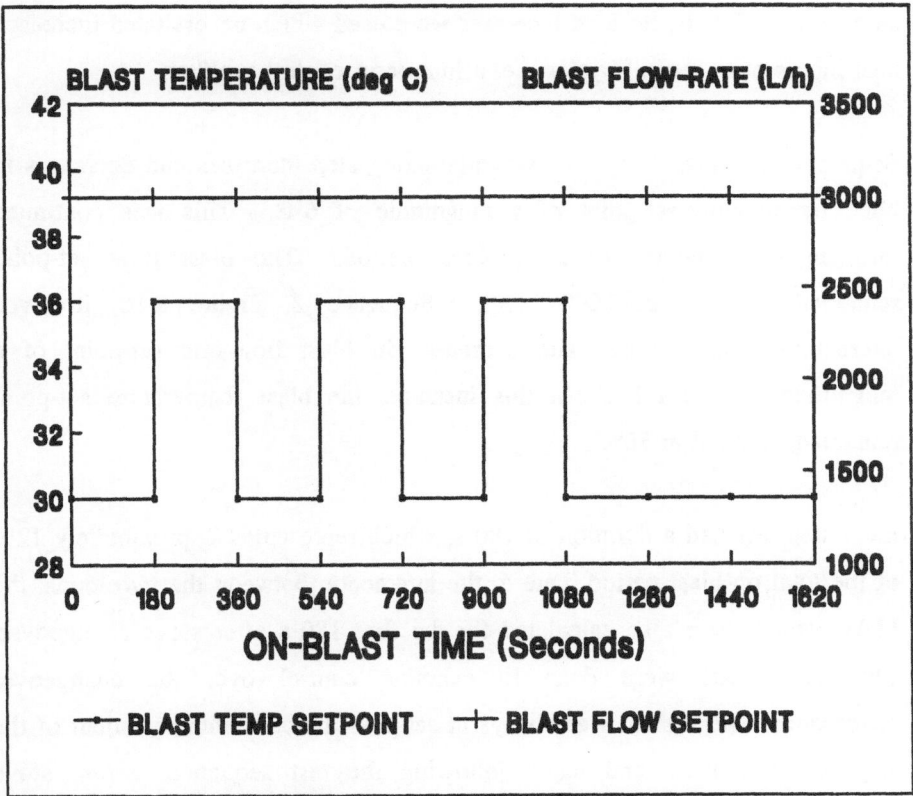

Figure 8.15 Step test sequence 1, load changes in blast temperature

quantitatively the controller performances. A lower ITAE value was the indication of better control. The behaviour of the controllers over the stove changeover transient was also examined.

The step changes in the set-points for the blast temperature and the blast flow-rate were chosen to be approximately 10-15% of the operating range. These load changes were centred within the operating range. Load increases were of a magnitude which could be supplied by the on-blast stove in the short term. That is, no load increases were used which necessitated increased heat inputs over several cycles, i.e. a long term control problem.

Sequence 1, Figure 8.15, involved alternating step increases and decreases in blast temperature set-point of a magnitude of 6°C. This was continued throughout the course of an on-blast period. The blast flow set-point remained constant at 3000 l h^{-1}. Sequence 2, Figure 8.16, involved alternating step increases and decreases in blast flow-rate set-point of a magnitude of 600 l h^{-1}. In this instance, the blast temperature set-point remained constant at 30°C.

Each step test had a duration of 180 s, which represented approximately 12% of the total on-blast period. Due to the interaction between the two loops, the ITAE values were also calculated for the first 180 s after stove changeover. The latter tests were done to examine control over the changeover disturbance. The system was always at set-point before commencement of the step test sequences and again following the test sequence, before stove changeover. ITAE values were obtained for both stove 1 and stove 2 in on-blast phase. Consistent results were obtained for both stoves.

The typical test procedure involved operating the two-stove system through several cycles of on-blast and on-gas phases. Blast temperature and flow-rate were controlled by either the GMC or base-case PID control scheme. Test conditions were maintained as consistent as possible between the different experimental runs to eliminate spurious effects on controller performance. The air supply was regulated at 280 kPa gauge. The on-gas manual valve was fully opened, giving constant on-gas flow of approximately 5000 l h^{-1}. The heater set-point was set at 90°C for all tests. Where possible, experiments were performed with consistent ambient conditions. At the end

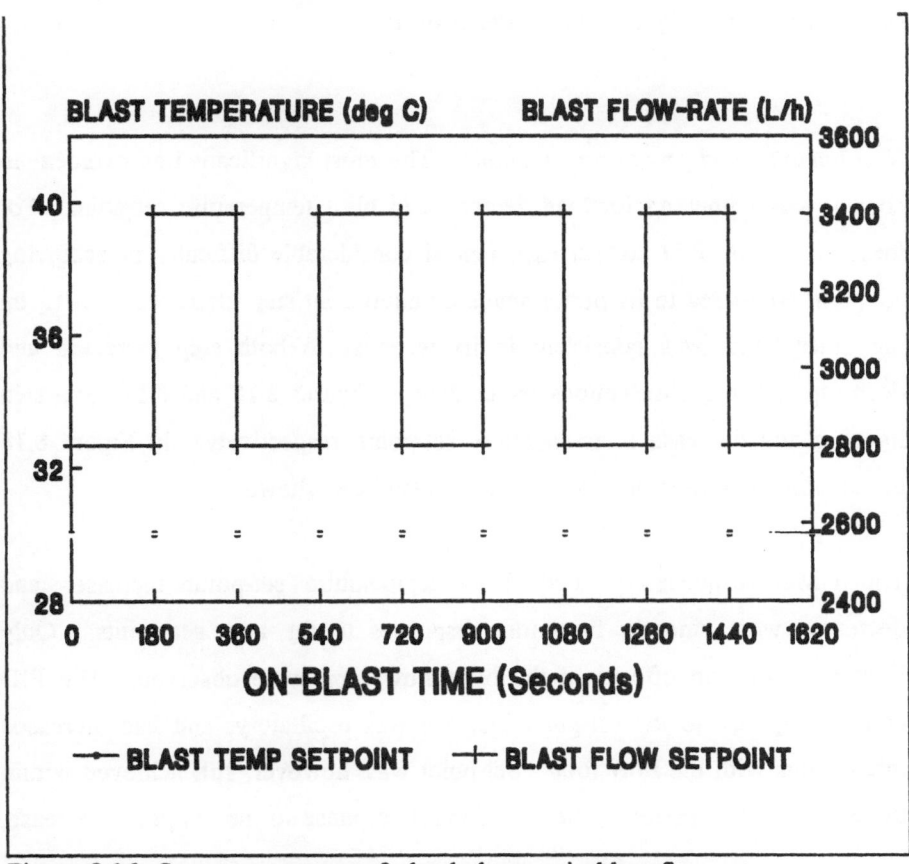

Figure 8.16 Step test sequence 2, load changes in blast flow-rate

of the prescribed step test sequence, the stoves exchanged duties. The new on-blast stove was then placed through the same sequence of load changes. This procedure was repeated for several cycles.

(a)   Load changes in blast temperature (sequence 1)

Figure 8.17 presents the ITAE values for load changes in blast temperature for both GMC and PID control. In all cases, GMC performed at least as well as, and frequently significantly better than, PID.

Non-linearity and process interactions. The most significant improvement in control was witnessed for load decreases in blast temperature set-point. For these situations, PID control experienced considerable difficulty in achieving set-point compared to its performance on equivalent step increases. GMC, on the other hand, was consistent in its response to both step increases and decreases. These observations are evident in Figures 8.18 and 8.19 for a step increase and decrease in temperature set-point, respectively. In Figure 8.18 the manipulated variable responses have also been shown.

The GMC responses to both blast temperature set-point increases and decreases were smooth, first-order responses to the new set-points. Only limited, short-term effects on the blast flow loop were observed. The PID control response to the set-point increase was oscillatory, and had increased interactions with the flow loop. Set-point was, however, still achieved within the 3 minute test period. The PID control response to the set-point decrease,

218

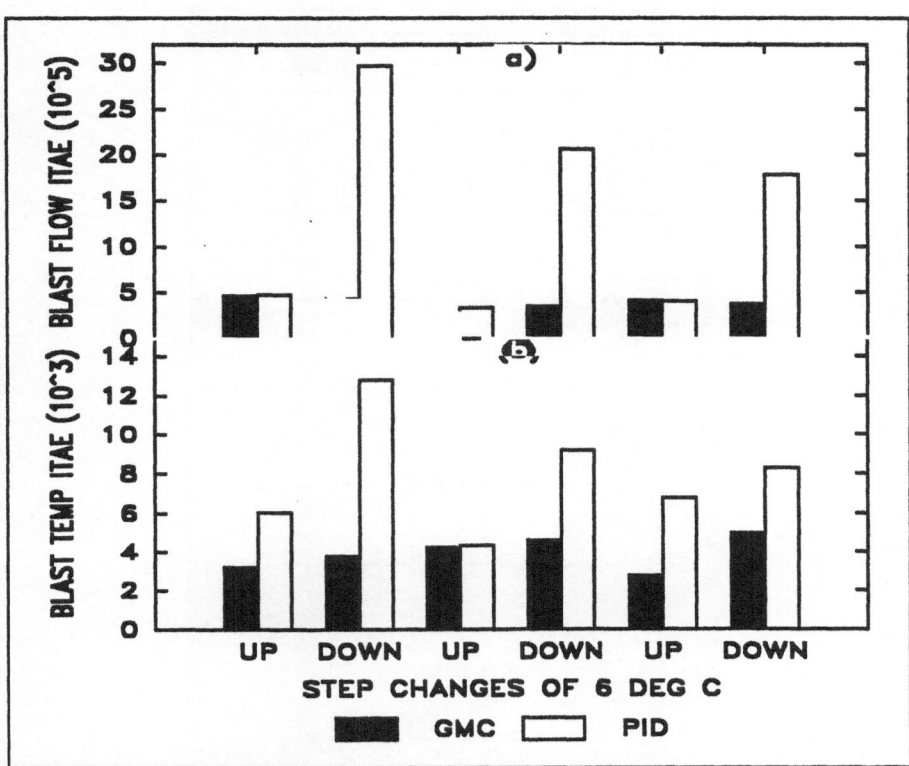

Figure 8.17 Controller ITAE values for blast temperature set-point sequence (sequence 1)

in marked contrast, showed considerably increased oscillation in blast temperature and severe interactions with the blast flow. In fact, the PID controller did not achieve the set-point decrease within the test period.

The difference in process behaviour for step increases and decreases in blast temperature represents a serious system non-linearity which the PID controllers were ill equipped to handle. Although coping satisfactorily with a

219

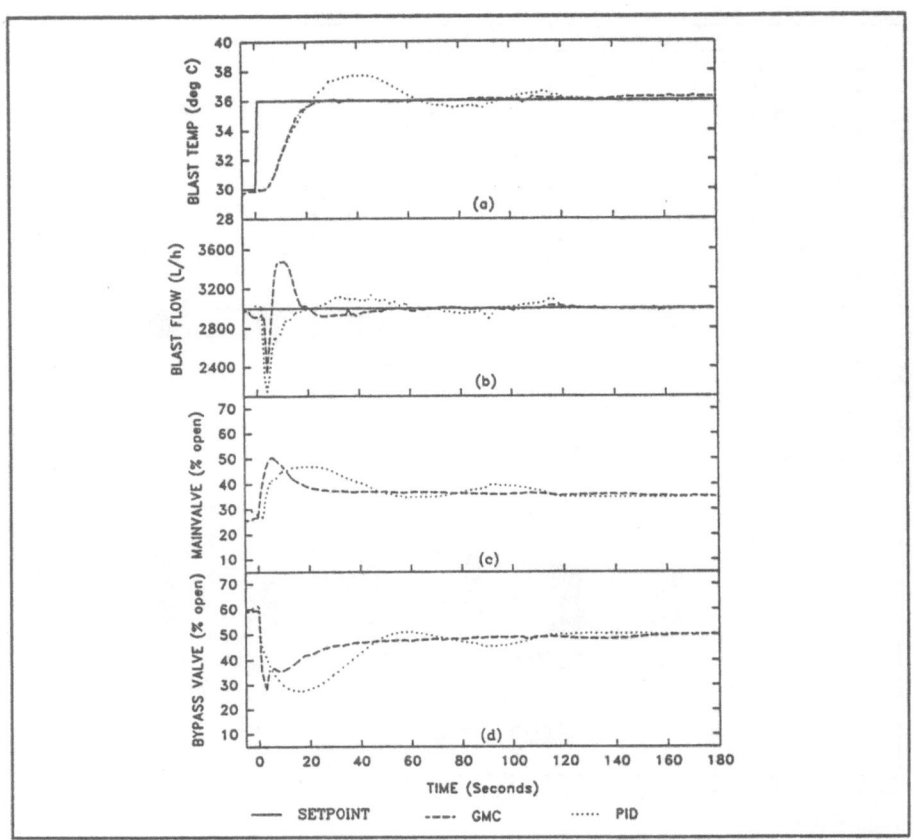

Figure 8.18   Blast temperature set-point step increase from 30°C to 36°C (step 1, sequence 1)

step increase, the same PID tuning constants caused excessive control action for the equivalent step decrease. Detuning of the PID controllers would lessen the control action. It would, however, simultaneously deteriorate performance for other load changes. Furthermore, oscillation between the two PID control loops for blast temperature and flow emphasized the strong interactions existing between these control loops. The PID controllers could not compensate for these interactions.

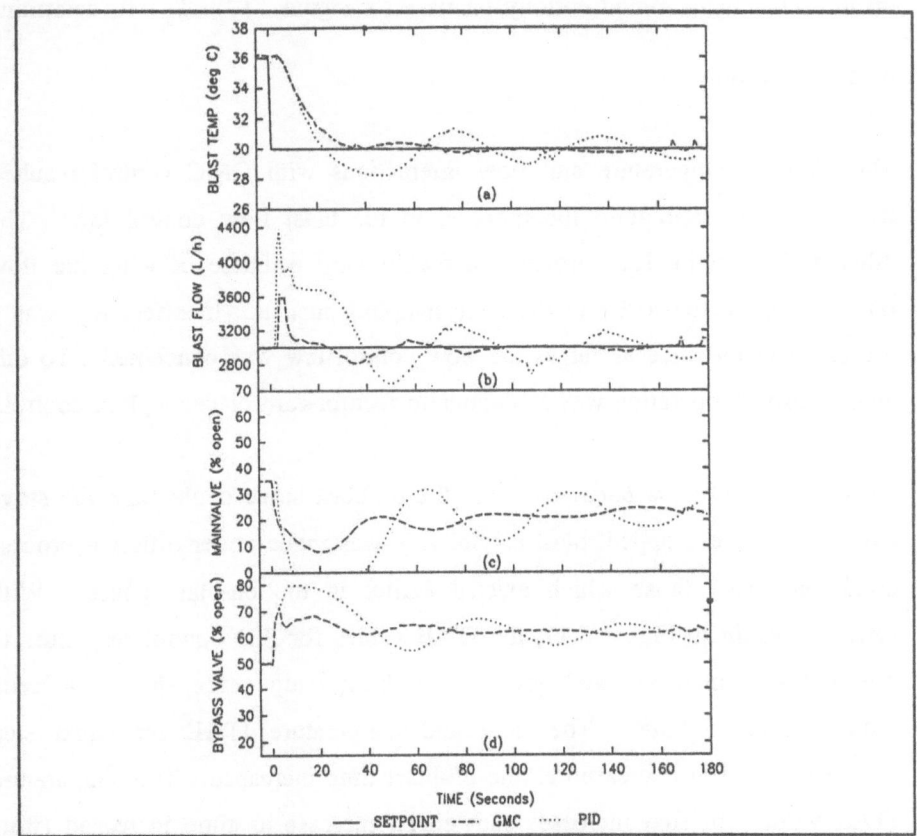

Figure 8.19   Blast temperature set-point step decrease from 36° to 30 °C (step 2, sequence 1)

However, the GMC controller was able to give consistently good responses for both temperature load increases and decreases, while significantly reducing interactions. The good response to temperature load change may be attributed to the steady state model approximation to the required flow ratio for the current set-point and stove condition. The traditional PID controller, on the other hand, relied entirely on its error signal $T_3^* - T_{3_k}$ to determine its control action.

The reduced temperature and flow interactions with GMC control resulted from the flow ratio term incorporated in the blast flow control law. The 'filtered' flow ratio, R_{filt}, provided a feedforward estimate of what the flow ratio was expected to be in the next sampling instant. In effect R_{filt} was a measured disturbance as far as the flow control law was concerned. To this end, controller operation was analogous to feedforward or decoupling control.

Time-varying process behaviour. As the on-blast stove cooled and the stove exit temperature dropped, blast air delivery was made under different process conditions from those which existed earlier in the on-blast phase. With reference again to Figure 8.17, the ITAE values for PID control responses to repeated step increases and decreases in blast temperature showed a trend with respect to time. The flow and temperature ITAE for equal step decreases actually decreased as the on-blast time increased. The temperature ITAE values for step increases showed an increase as time increased (flow ITAEs for the latter showed no significant change). The GMC controller, however, was essentially consistent in its ITAE values throughout the course of the on-blast period and for both step increases and decreases.

Figure 8.20 shows the GMC and PID control responses for a step decrease in blast temperature near the end of the on-blast period (step 6 in sequence 1).

The GMC response was similar to its response for the earlier step decrease (see Figure 8.19). The PID control response, however, was improved compared to its response to the earlier step decrease. For the later step decrease, set-point was actually approached within the 3 min test period, unlike the earlier step decrease.

The GMC controller, in marked contrast to the PID controllers, showed fairly consistent control performance over time. This was again directly attributable to the incorporation of the process model in the control law.

(b)   Load changes in blast flow-rate (sequence 2)

Figure 8.21 presents the blast flow and temperature ITAE values for GMC and PID control of the rig under step test sequence 2 (see Figure 8.16). As in the case of the blast temperature sequence, the GMC control again performed at least as well as, and generally significantly better than, PID control.

Time varying behaviour and process interactions.   Figure 8.21 shows an obvious trend in PID control performance over the course of the on-blast phase. At the outset, the PID controllers demonstrated extremely poor control, as exhibited by the very high ITAE values for both blast temperature and flow-rate. Performance, however, improved steadily, essentially becoming equivalent to that of GMC at the very end of the period.

The GMC controller responses, again in marked contrast, were consistently good throughout. No significant variation over the on-blast period was exhibited.

223

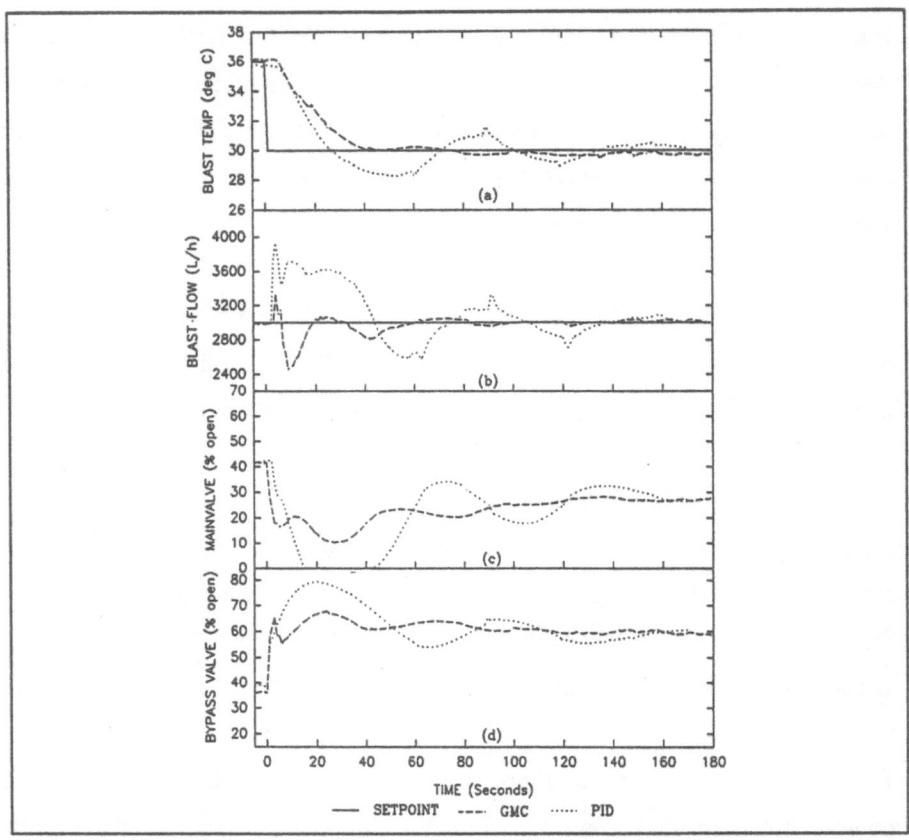

Figure 8.20   Blast temperature set-point step decrease from 36° to 30°C late in on-blast phase (step 6, sequence 1)

Figure 8.22 presents the GMC and PID control responses to the 600 lh^{-1} step increase in blast flow, at the beginning of an on-blast period (step 1, sequence 2).   The GMC response was rapid, with little overshoot or oscillation in the blast flow.   Minimal effect on the blast temperature response was observed.   The PID control response, however, showed considerable oscillation in the blast flow-rate and significant interaction

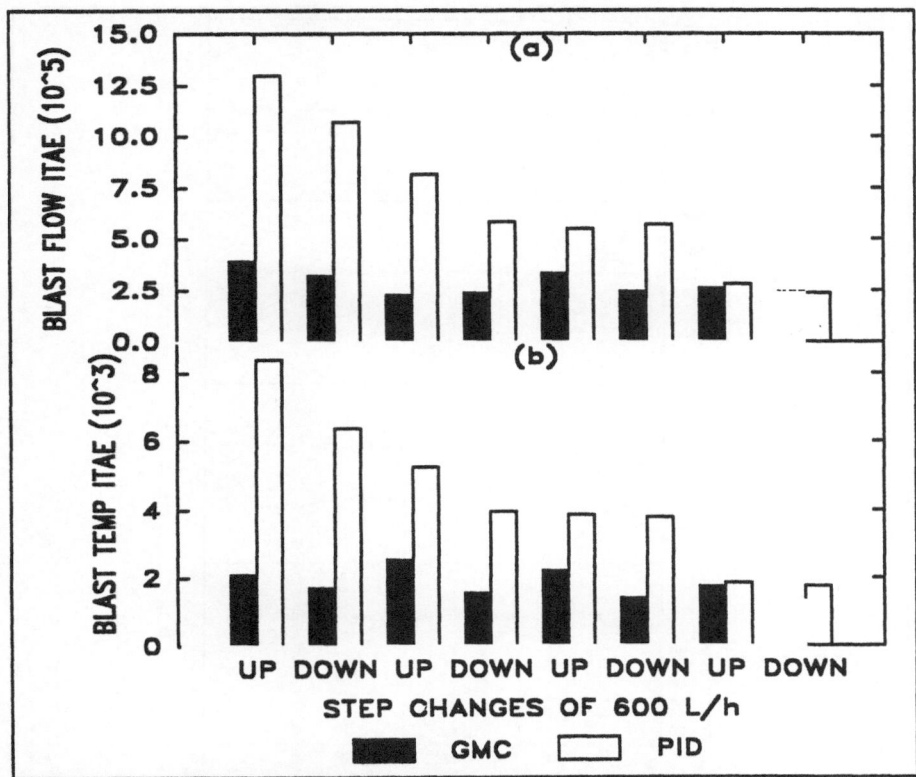

Figure 8.21   Controller ITAE values for blast flow-rate set-point sequence (sequence 2)

between the flow and temperature loops.

Figure 8.23 shows GMC and PID control responses to the last 600 l h^{-1} step increase in blast flow set-point (step 7, sequence 2).   This step change occurred over 1000 s after the first step increase discussed above.   The GMC control response was very similar to that of the earlier step increase (compare with Figure 8.22) and was again very good.   The PID controller response was

225

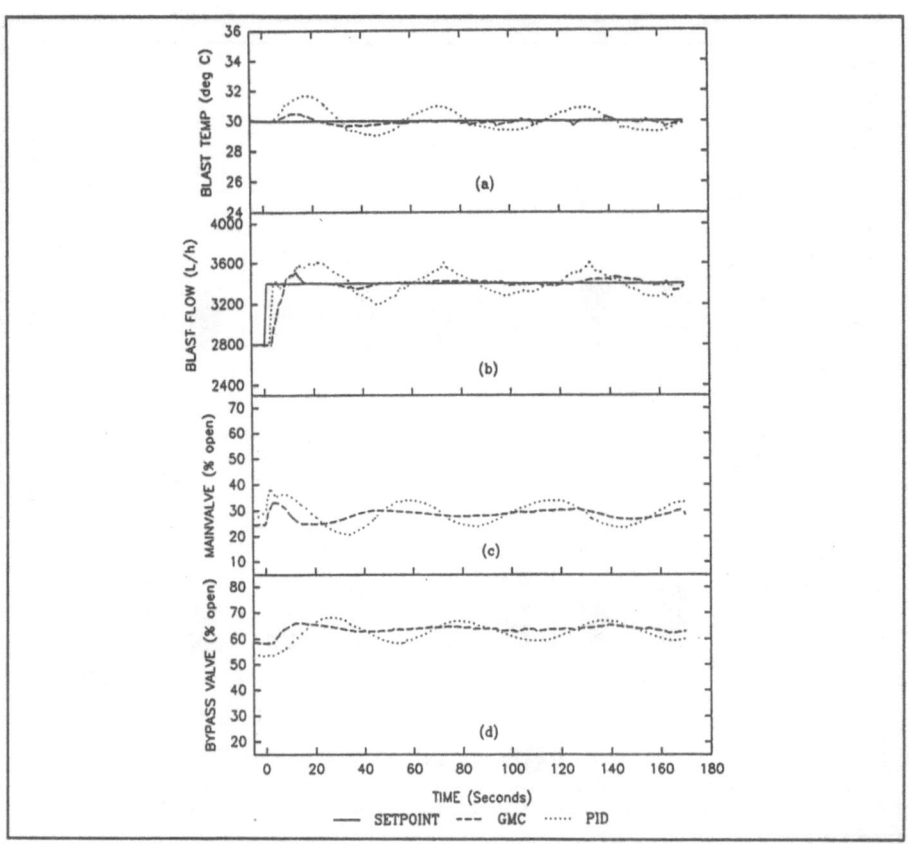

Figure 8.22 Blast flow-rate set-point step increase from 2800 to 3400 1 h^{-1} late in on-blast phase (step 1, sequence 2)

dramatically improved over its earlier performance. Set-point was achieved more quickly, with little effect on the blast temperature loop.

Linear PID control was unable to adapt to the time-varying process conditions. The model-based GMC controller did accommodate this behaviour, and therefore supplied improved responses throughout. This

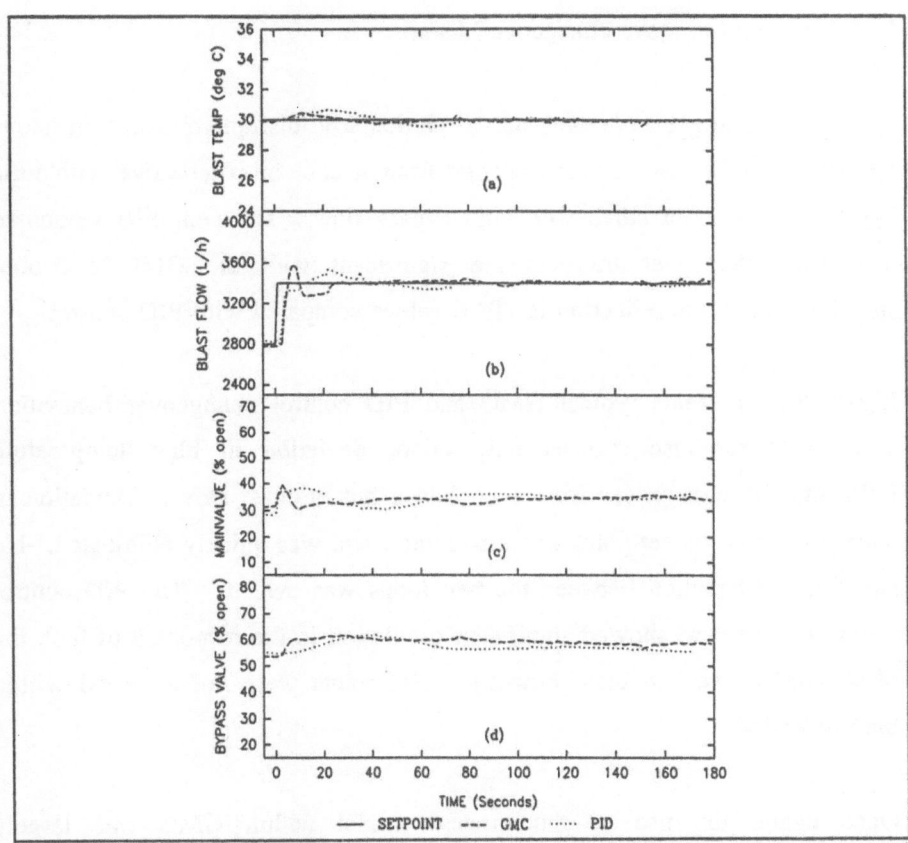

Figure 8.23   Blast flow-rate set-point step increase from 2800 to 3400 l h^{-1} late in on-blast phase (step 7, sequence 2)

improvement resulted largely from reduced interactions with the blast temperature loop, as discussed earlier.

227

(c)   Responses to stove changeover disturbance.

Stove changeover represents a particularly disruptive event in stove operation.   There is a sudden change from a cool on-blast stove with little bypass air, to a hot stove with high bypass flow.   The controller responses over the changeover transient are significant with the GMC controller showing a dramatic reduction in ITAE values compared with PID control.

Figure 8.24 presents typical GMC and PID control changeover behaviour. The GMC response showed only minor deviation in blast temperature following the changeover from a cool to a hot on-blast stove.   Deviation in blast flow from its set-point was also minor, and was quickly eliminated.   No significant interaction between the two loops was evident.   The PID control response, however, showed significant oscillation in the responses of both the blast temperature and blast flow-rate.   Set-points were not achieved within the test period.

Once again, the use of the process model within GMC was largely responsible for the improved control.   Stove changeover and the subsequent increase in stove exit temperature, T_2, represented disturbances to the system. Feedforward compensation minimized the effect of changeover on the blast temperature.   The base-case PID control response had to wait until the effect on the blast temperature, T_3, had become evident, and only then did it react.

The process interactions between the blast temperature and flow-rate were again reduced with GMC by the use of the flow ratio, R_{filt}, in the blast flow

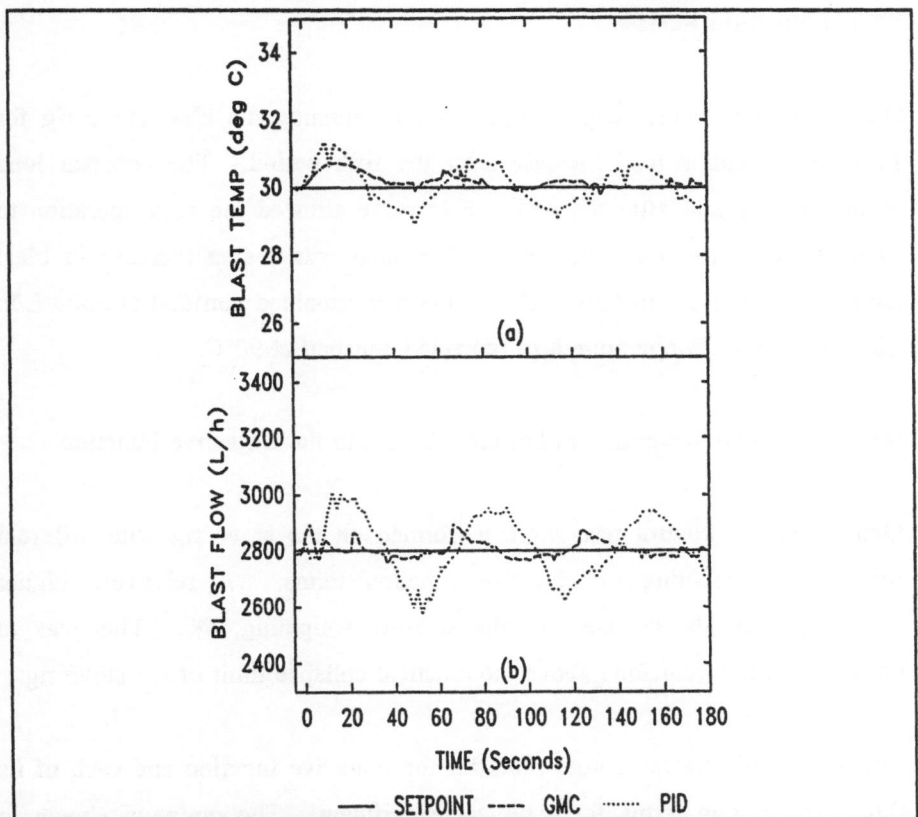

Figure 8.24 Controller responses to stove changeover

control law. As observed in the temperature set-point step changes (sequence 1), the control actions of the blast temperature controller were disturbance inputs to the blast flow controller. Feedforward compensation was made to maintain the blast flow set-point and minimize oscillation.

8.5.2 Long-term Results

Online optimal control experiments involved running the blast stove rig for 15 periods, with a load increase after the fifth period. The constant load periods before and after the period 5 increase allowed the rig's operation to stabilise out. The load increase implemented was a step increase in blast temperature from 36 to 40°C. Blast flow-rate remained constant at 3500 L/h. The on-gas inlet temperature also remained constant at 90°C.

(a) Preferential Weighting of Different Terms in the Objective Function.

Online optimal control tests were performed on the stove rig with different preferential weighting of objective function terms. A relatively higher weighting was always used for the stability weighting, W_s. This was to ensure operation remained above the practical collapse limit of the stove rig.

Figure 8.25 illustrates contour plots for the objective function and each of its three contributing terms for a typical experiment. The optimum shown in Figure 8.25 (a) was determined during the online optimisation of rig operation for the period 5 load increase. The optimisation was performed based on the stove conditions at the end of period 4 and the desired load for period 5. The optimum defines the setpoints to be delivered during period 5. The weighting factors were specified as:

Thermal Efficiency Weighting,	$W_t = 1.0$
Cyclic Stability Weighting,	$W_s = 100$
Load Response Weighting,	$W_f = 5.0$

In the region of the optimum in Figure 8.25 the contributions by each of the three terms are of approximately the same order of magnitude. As the on-gas heat input is decreased and the collapse point is approached the objective function becomes dominated by the cyclic stability term. This is apparent by the increasing slope of the functions in Figure 8.25 (a) and (c). In each case the effect of the quadratic form of the objective function is obvious by the changing distance between contour lines.

The optimum point in Figure 8.25 (a) specifies a blast temperature below the desired load of 40°C. This was a consequence of the dominant thermal efficiency and cyclic stability terms, which prevented the full load from being delivered.

(b) Changing the Thermal Efficiency Weighting

Figure 8.26 shows the blast temperature and on-gas flow setpoints determined by the optimal control strategy for different values of the thermal efficiency weighting factor, W_t. The first experiment employed a value of 1.0, twice the size of that used in the second run. For both runs the other two weighting factors were $W_f = 5$ for response speed, and $W_s = 100$ for stability. Both runs showed similar responses in the blast temperature response, with the new load being fully delivered for the second period after the load

231

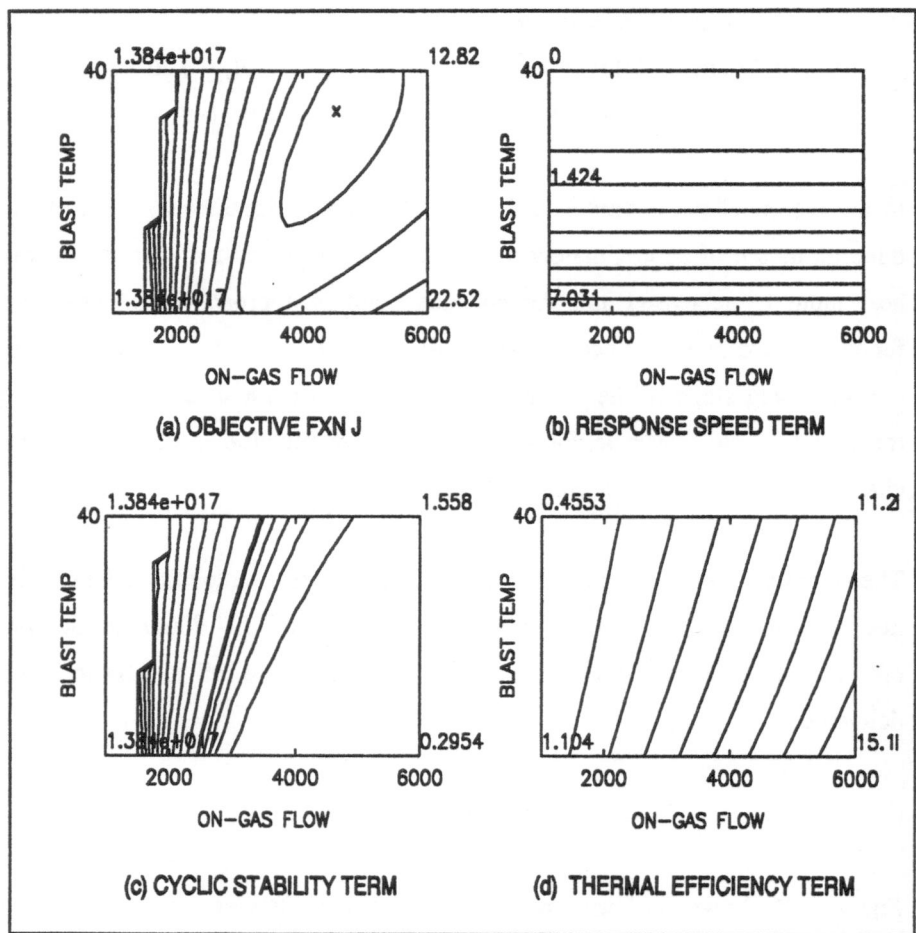

Figure 8.25 Contour Plots of the Objective Function and its three component terms at the period 5 load increase

change.   A marked difference, however, is evident in the heat input specifications made by the optimiser.   The first run, with the higher thermal efficiency weighting, attains an eventual heat input rate 10% lower than the other run.

232

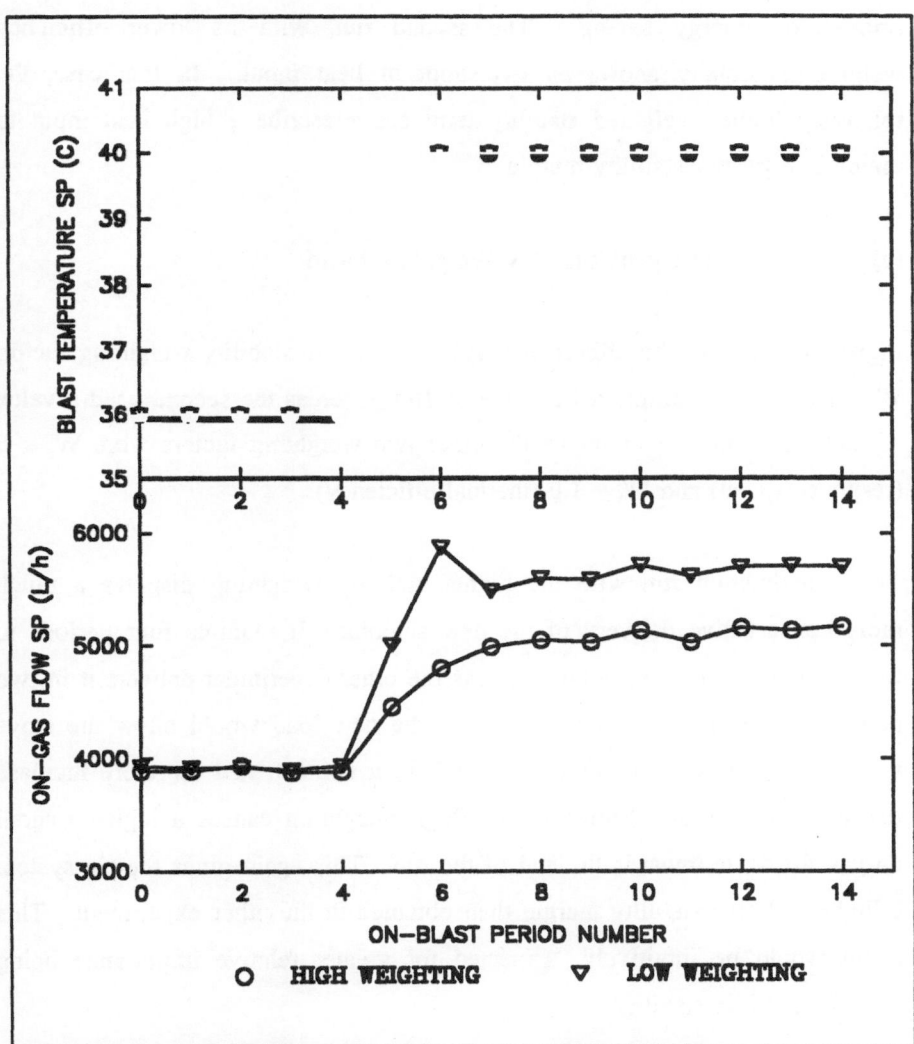

Figure 8.26 Effects of Changing the Thermal Efficiency Weighting, W_t

This result would be expected for greater relative importance being placed on
thermal efficiency. Even the overdamped response is indicative of a higher

233

priority on energy saving. The second run, with its lower efficiency weighting, actually shows an overshoot in heat input. In this case, the relatively higher weighted stability term can prescribe a high heat input to maintain a greater stability margin.

(c) Changing the Cyclic Stability Weighting Factor

Figure 8.27 shows the effects of varying the cyclic stability weighting factor, W_s. The first run employed a value of 100, whereas the second used a value of 150. For both experiments the other two weighting factors were $W_f = 5$ (response speed) and $W_t = 1.0$ (thermal efficiency).

The experimental run with the higher stability weighting displays a much more conservative delivery of the new setpoint. It requires four periods to fully deliver the new setpoint, whereas the other experiment delivers it in two periods. The more gradual delivery of the new load would allow the stove system extra time to increase its overall heat content, and therefore increase the stability margin. Moreover, the high weighting causes a higher overall on-gas flow-rate towards the end of the run. This again gives the rig system a higher relative stability margin than obtained in the other experiment. This result would be intuitively expected for greater relative importance being placed on cyclic stability.   .

(d) Changing the Response Speed Weighting Factor

Figure 8.28 shows the effects of varying the cyclic stability weighting factor,

W_f. The first experimental run employed a value of 10, and the second used 5. The common weighting factors were W_s = 150 and W_t = 1.0 for cyclic stability and thermal efficiency, respectively.

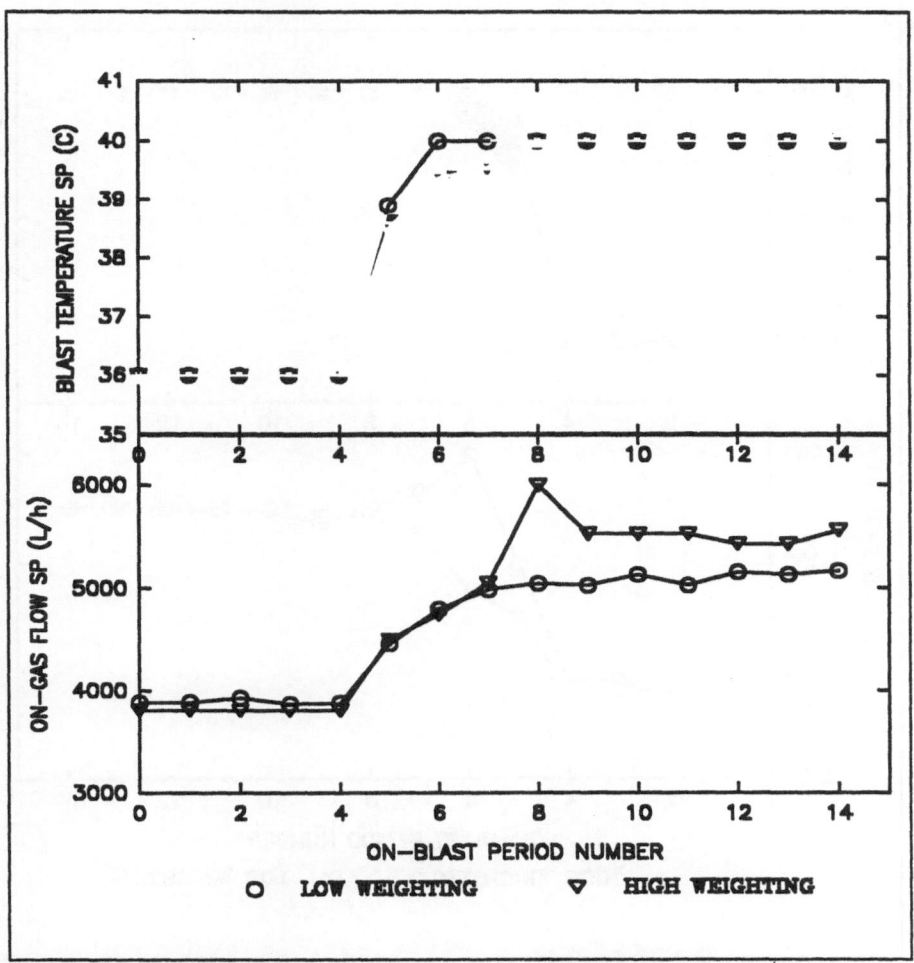

Figure 8.27 Effects of Changing the Cyclic Stability Weighting, W_s

The experimental run with the higher response speed weighting displays a much more rapid delivery of the new setpoint than the second run. The first

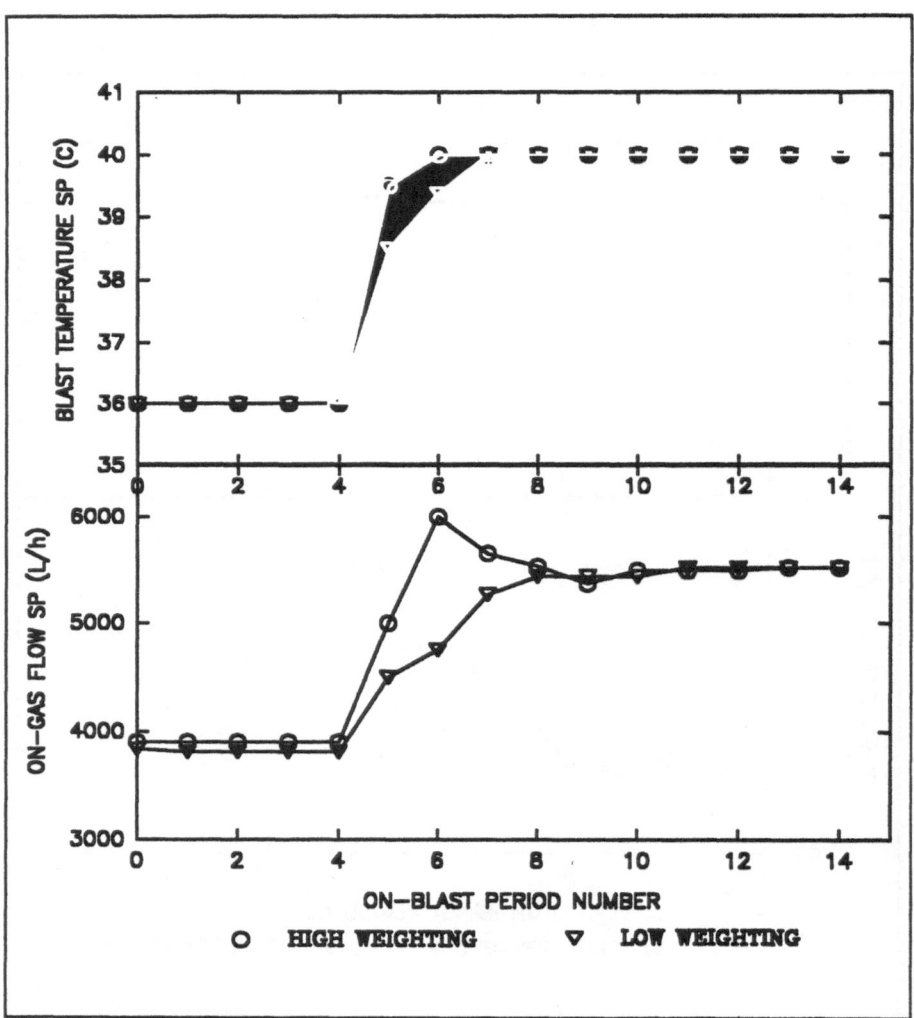

Figure 8.28 Effects of Changing the Control Response Weighting, W_f

run requires only two periods to fully deliver the new setpoint, whereas the second delivers it in three periods.

The more rapid control response due to the high weighting occurs with a much more liberal addition of on-gas heat to the rig system. In fact, the on-gas flow setpoint overshoots its eventual equilibrium value. Thermal efficiency was clearly compromised with the higher emphasis on rapid control response. The second experiment, in contrast, displays a much more conservative response both in blast temperature delivered and on-gas flow-rate supplied. Once again, this result would be intuitively expected when higher importance is placed on rapid control responses.

An interesting result was obtained when the response speed weighting was specified at a much lower value. Optimal control with the response weight $W_f = 1$ gave the results shown in Figure 8.29. At period 5, when the optimisation program was executed for the load increase, the low response weighting actually allowed the blast temperature to drop below its previous value. Clearly, the relatively higher stability and thermal efficiency weightings were dominant and drove the system in the direction opposite to the intended load change.

(e) Comparison to Conventional Stove Control

The conventional operating method for blast stoves in industry has long been the fixed-period method. The advantage of this fixed-period method solely lied in its simplicity. Strict control over the average heat input was not

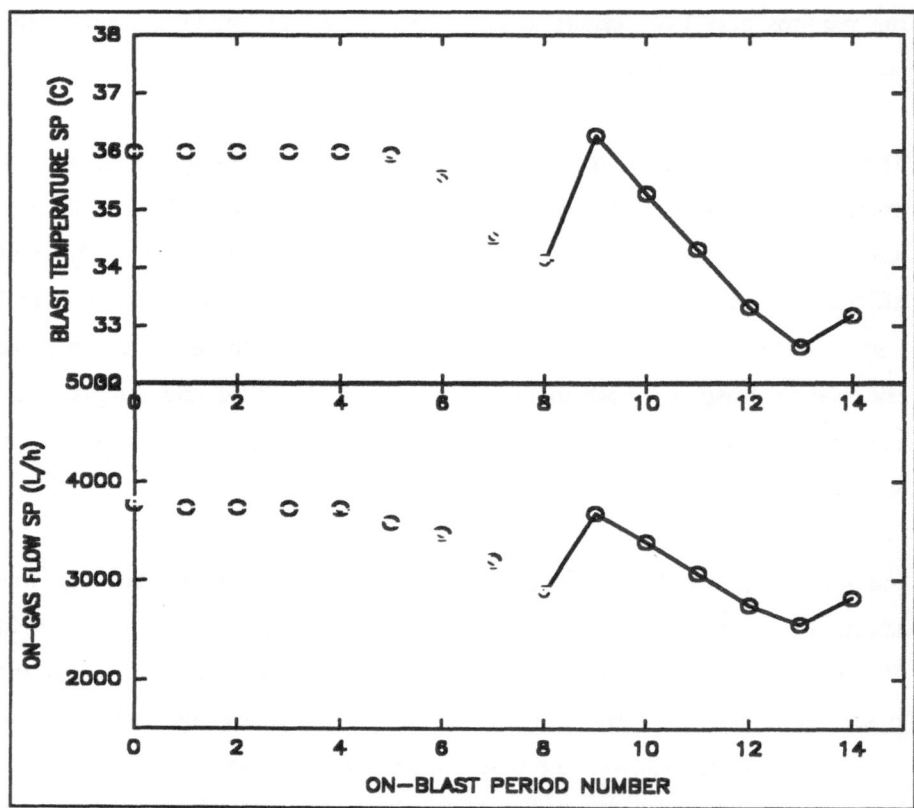

Figure 8.29 Effects of Very Low Control Response Weighting

necessary as a significant level of residual heat was maintained in the stove.
Similarly, no condition was placed on the final fraction of bypass mixing.
The operator could accommodate small load increases utilising this residual
heat. For larger load increases, a small increase in load would initially be
delivered, with subsequent heat increases in the next on-gas phases to
accommodate the remaining load increase.

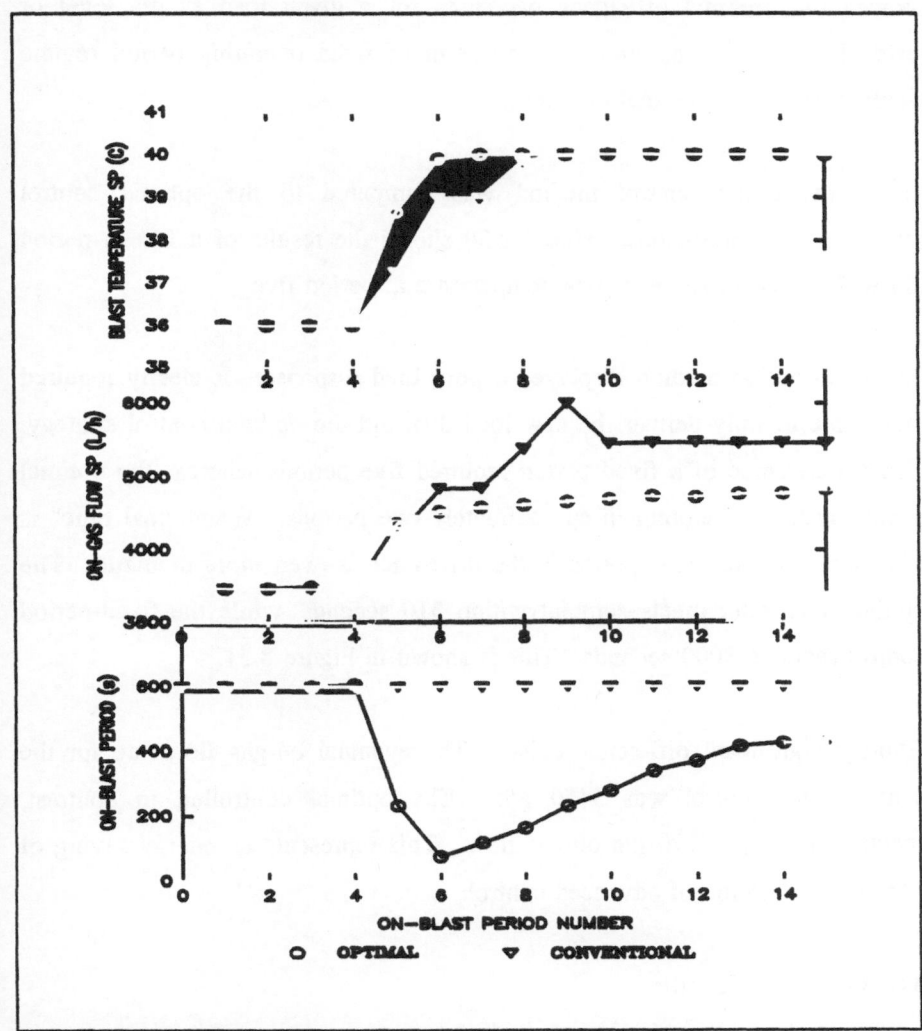

Figure 8.30 Optimal and Conventional (Fixed-Period) Control Performance - Period Number

The simplicity of this method, however, comes at a cost. Jeffreson (1979) showed that thermal efficiency decreases for a given load as the level of residual heat increases, and that any form of fixed operating period regime would show lower thermal efficiency.

The conventional control method was compared to the optimal control scheme using simulations. Figure 8.30 shows the results of a fifteen period run with a step increase in blast temperature at period five.

The fixed-period method displayed a poor load response. It clearly required more time to fully deliver the new load than did the optimal control strategy. The maintenance of a fixed period required five periods whereas the optimal control achieved setpoint in approximately two periods. When "real time" is considered, rather than "periods", the difference is even more dramatic. The optimal controller meets setpoint within 310 seconds, while the fixed-period control requires 3000 seconds. This is shown in Figure 8.31.

Another significant difference exists. The eventual on-gas flow-rate for the conventional control was 5450 g/h. The optimal controller, in contrast, specified a frugal 4770 g/h on-gas flow. This represents an energy saving of 14% by application of advanced control.

8.6 Economic Benefits

The economic benefits of implementing the long term controller were suggested in the previous section. The long term controller made a 14%

240

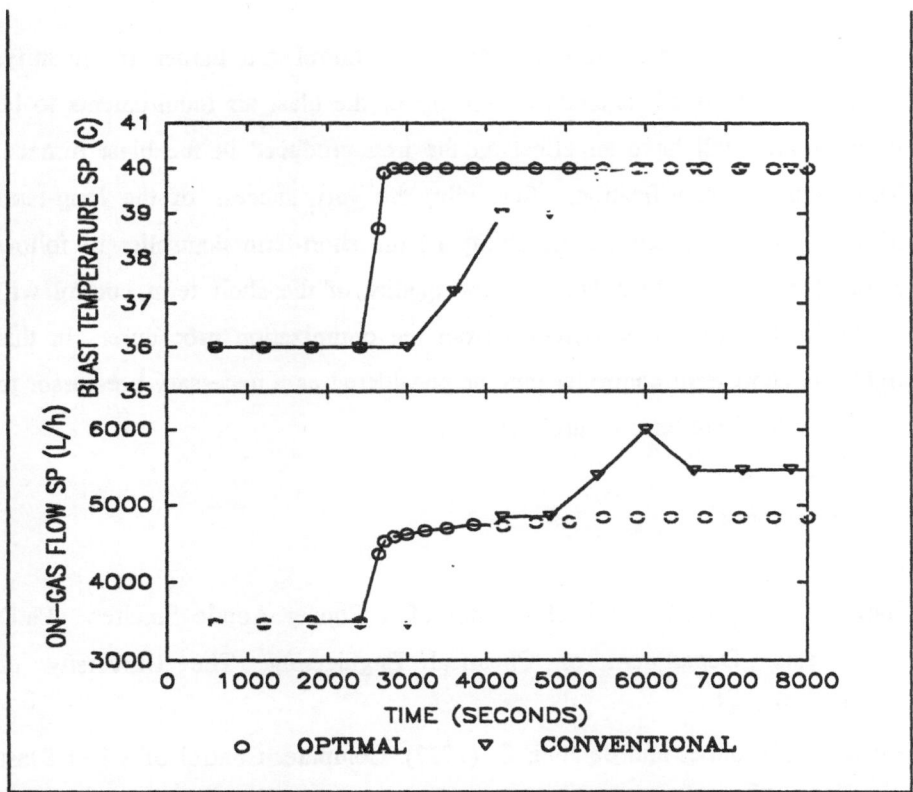

Figure 8.31   Optimal and Conventional (Fixed-Period) Control Performance - Real Time

reduction in energy usage and an order of magnitude increase in the speed of response. Of course, these benefits have been demonstrated on the pilot scale apparatus and require future verification on full scale stoves. The economic value associated with these improvements requires knowledge of the cost of energy, and on the effect of quicker delivery of blast air at the specified temperature and flowrate has on the iron quality and quality produced in the

241

blast furnace. These factors are likely to be site specific.

The benefits of the improved short term control are harder to quantify. Again, the effects of "smoother" delivery of the blast air requirements to be blast furnace will have an effect on the iron produced in the blast furnace. This requires quantification. Secondly, the very success of the long-term controller is dependent on the ability of the short-term controller to follow setpoint changes. Degradation in the quality of the short term control will erode the benefits to be obtained from the optimization procedure. In this light, the short term controller may be considered as a necessary pre-cursor to achieving the long term control benefits.

8.7 References

Bateman I. (1992) Optimal Control of a Copper Anode Smelter. Ph.D Thesis, Department of Chemical Engineering, The University of Queensland.

Beets J., Elshout J. and de Jong G. (1977). Computer Control of a Hot Blast Stove System. Journal A (Belgium) 18,1:31-37

Box M.J. (1965) A New Method of Constrained Optimization and a Comparison with Other Methods. Computer Journal 8:42-52

Jeffreson C.P. (1979) Feedforward Control of Blast Furnace Stoves. Automatica Vol 15:149-159

Mitter G., Delavos G. and Conert W. (1981) Computer Control and Optimization of a Blast Furnace Stove Operation. Iron and Steel Engineer 58,10:43-50

Nose K., Takemura M. and Morita T. (1984) Systems Engineering Approach to the Optimisation of Hot Blast Stoves. IFAC Proc. Series 1984 5:47-54

Smith C.A. and Corripio A.B. (1985) Principles and Practice of Automatic Process Control. Wiley

Turnbull Control Systems (1988) Reference Manual. Sussex, England.

CHAPTER 9

CONCLUSIONS

One of the aims of this monograph was to illustrate, through a number of industrial problems, the benefits of applying modern model-based control. The examples presented in this work have clearly demonstrated the applicability of one such modern control strategy, namely Generic Model Control (GMC). While it can always be argued that a particular problem brings a unique set of circumstances, this work illustrates that a wide variety of different control problems can be "solved" using a flexible control algorithm. It hopefully does meet industrial practitioners' needs of providing reference examples to give others confidence that modern "theory" can indeed be successfully applied.

The monograph has also illustrated implementation issues that have to be considered to arrive at a successful conclusion. The narrative style of chapter seven clearly illustrates that at times perseverance and sometimes faith is required. Rarely does an application work the first time. Chapter three demonstrates that the 90:10 principle can also apply - 90% of the return for 10% of the effort. Further refinement of the results obtained are probably possible, but are they justifiable?

The degree of sophistication of implementation hardware and software required to implement GMC has been shown to be modest. The pH controller of chapter three was implemented in an on-line computer, albeit a small machine, using standard commercial software. The blast furnace stove controller of chapter eight, and the distillation column controller of chapter four were implemented using PC computers, while the extruder controller of chapter six used a commercial PLC.

Operator acceptance of advanced control schemes is very important. Chapter four discussed this point and illustrated many practical considerations required to obtain this vital component. The ability of the DRYER MASTER controller, described in chapter five, to achieve operator acceptance can be inferred from the commercial success of the company in marketing this product.

Each individual case study has highlighted the superior performance of the model based controller. Substantial benefits have been illustrated, with a payback time of six months for the dryer controller of chapter five, $US250,000 to $US2,000,000 for examples in distillation control in chapter four, and $CDN2,000,000 for the reactor control described in chapter seven. These are substantial savings and clearly the companies involved have reaped large rewards.

One of the advantages of GMC illustrated through this monograph is its ability to use a variety of model types. Chapter seven used an empirical model, chapter four illustrated semi-empirical models based on approximate

246

dynamic representations and semi-rigorous steady-state descriptions, while chapter 8 took a more fundamental approach and derived a model from first principles. The ability to use a variety of model types, which all must contain various degrees of inaccuracies, improves the flexibility of the method and enhances its usability.

The monograph has also illustrated three other issues. Firstly, the ability to control processes in the presence of constraints was illustrated in chapters four and eight. In particular, chapter eight showed a clear method of not only incorporating traditional bound constraints on both the manipulated and the controlled variables, but also functional constraints representing higher level operating objectives. Secondly, controlling processes that incorporate dead-times was illustrated in chapter seven. Finally, the flexible specification of the desired closed loop performance was demonstrated. Chapter 2 and other chapters illustrated a time-domain approach while chapter three illustrated a closed-loop pole placement strategy.

This work has highlighted the need for process models to obtain good process control. Despite the inherent properties of GMC, ultimately good performance is dependent on a good model. A modelling methodology that assists practitioners in obtaining reliable process models would greatly advance the practice of modern control theory.

Finally this monograph was not meant to be a comparison of different modern control methods, concluding that technique A was better than technique B. It was meant, as has been stated earlier, to provide some further reference examples for industrial practitioners to "take heart and sally forth". Others may care to provide detailed comparisons.

INDEX